Spatial Microsimulation: A Reference Guide for Users

Understanding Population Trends and Processes

Volume 6

Series Editor

J. Stillwell

In western Europe and other developed parts of the world, there are some very significant demographic processes taking place at the individual, household, community and national scales including the ageing of the population, the delay in childbearing, the rise in childlessness, the increase in divorce, the fall in marriage rates, the increase in cohabitation, the increase in mixed marriages, the change in household structures, the rise in step-parenting and the appearance of new streams of migration taking place both within and between countries. The relationships between demographic change, international migration, labour and housing market dynamics, care provision and intergenerational attitudes are complex to understand and yet it is vital to quantify the trends and to understand the processes. Similarly, it is critical to appreciate what the policy consequences are for the trends and processes that have become apparent. This series has its roots in understanding and analysing these trends and processes.

This series will be of interest to a wide range of individuals concerned with demographic and social change, including demographers, population geographers, sociologists, economists, political scientists, epidemiologists and health researchers as well as practitioners and commentators across the social sciences.

For further volumes:
http://www.springer.com/series/8113

Robert Tanton • Kimberley L. Edwards
Editors

Spatial Microsimulation: A Reference Guide for Users

Springer

Editors
Robert Tanton
The National Centre for Social
and Economic Modelling (NATSEM)
University of Canberra
Canberra, ACT, Australia

Kimberley L. Edwards
Centre for Sports Medicine
University of Nottingham
Nottingham, UK

ISBN 978-94-007-4622-0 ISBN 978-94-007-4623-7 (eBook)
DOI 10.1007/978-94-007-4623-7
Springer Dordrecht Heidelberg New York London

Library of Congress Control Number: 2012949355

Printed on acid-free paper

Springer is part of Springer Science+Business Media (www.springer.com)

Foreword

It is clear from the chapters of this book that the volume and quality of research on spatial microsimulation is on a rising curve—and so it should be because it is potentially the key to much of the future of urban and regional modelling. This is for two reasons: first, space is key to the modelling enterprise, and second, almost any other approach is bedevilled by the problems of high dimensionality—arrays with large numbers of subscripts and superscripts most of whose elements are zero. At the outset, therefore, microsimulation offers an efficient way of storing the complexities of cities and regions, reducing the storage space needed by several orders of magnitude. There is also a third reason: much government policy is enacted at a national level, and time after time, the consequences of such policies at a finer spatial scale are not investigated. Microsimulation has already shown that it can contribute enormously in this area.

This is not the place to summarise the book: that is done very effectively in the first and final chapters. What I can do is echo part of the argument of the authors of the concluding chapter and remark on the ongoing research agenda of which this book will be a foundation. The early chapters, which provide a blueprint on how to build a spatial microsimulation model, demonstrate another virtue: that such a model can integrate disparate data sources, often with different spatial and sector classifications. At present, there is often a specificity of purpose in these endeavours for example, to generate small area income distributions which are invaluable and which cannot be obtained in any other way. However, this suggests a first major research challenge: to be systematic about this and to see spatial microsimulation as the foundation of a comprehensive intelligence system—that has transformed disparate data into such a system. There are at least two ways in which this challenge can be approached. Perhaps not surprisingly, the mention of iterative proportional fitting and 'constraint variables' leads me to believe that there may be an entropy maximising version of the general problem which might enhance the deterministic route. Alternatively, I suspect there is a challenge yet to be fully taken on to articulate all the conditional probability distributions that underpin the stochastic route to microsimulation: there will be circularities here, and it will need some clever theoretical research to handle these.

A second challenge is reflected in the second half of the book: to make the models fully dynamic. Much progress has been made and is shown here, but a related challenge arises from one of the pleas of the authors of Chap. 16: to connect dynamic spatial microsimulation models to other urban models. This would mean representing 'players' other than the population—economic agents such as retailers, for example— and this would draw the field into the areas of modelling nonlinear systems with all the difficulties of path dependence and phase changes. These would generate discrete jumps in population behaviour—assuming that spatial interaction is fully built into the population models.

Many of the authors in this book have been members of a relatively small community that has driven spatial microsimulation forward. I can only applaud that commitment and now recognise that their efforts will surely be rewarded with the continuing growth of the field. This book will be an important agent in stimulating that growth.

London Alan Wilson

Contents

Part I
Background

Chapter 1
Introduction to Spatial Microsimulation: History, Methods and Applications

Robert Tanton and Kimberley L. Edwards

1.1 Introduction

Microsimulation as a method has been used in the social sciences since the pioneering work by Guy Orcutt and his colleagues (Orcutt et al. 1961). Several authors have extended the original work of Orcutt (Sutherland 1995; Orcutt and Glazer 1980; Zaidi et al. 2009), and the methodology has been incorporated into tools that can be used to examine the effects of policies before they are implemented, for example. The basic premise of microsimulation is that a more realistic picture of aggregate behaviour can be derived from looking at individual behaviour and modelling the interaction between the individual units in the system under consideration.

So far, much of this modelling has been applied at a national scale – such as the impact of tax changes on national income or the impact of health policy changes on the population. Recent advances in microsimulation by geographers have added a spatial dimension to these results. In the last few years, the number of spatial methods has expanded, and therefore this book is intended to bring together all of this recent research in what is now known as *spatial microsimulation*.

This book is intended to be a guidebook for practitioners looking to learn how to develop a spatial microsimulation model. The chapters show what sort of data are required, how to prepare these data, how different types of models have been developed using different methods and the limitations of each type of model, how to validate a model and finally, what the future is for spatial microsimulation.

The models are split into two types: static spatial microsimulation models and dynamic spatial microsimulation models. Static spatial microsimulation models use

R. Tanton (✉)
National Centre for Social and Economic Modelling, University of Canberra, Canberra, Australia
e-mail: Robert.tanton@natsem.canberra.edu.au

K.L. Edwards
School of Clinical Sciences, University of Nottingham, Nottingham, UK
e-mail: Kimberley.edwards@nottingham.ac.uk

R. Tanton and K.L. Edwards (eds.), *Spatial Microsimulation: A Reference Guide for Users*, Understanding Population Trends and Processes 6, DOI 10.1007/978-94-007-4623-7_1,

a static ageing technique which ages variables by either uprating them (for financial data) or reweighting them to future populations (for demographic data). Any policy change modelled is applied to the population, but the demographic processes of ageing, being born and dying, are not modelled. Static spatial microsimulation models are best for 'next day' analyses – for example, if we change this now, how will it impact the population tomorrow?

In comparison, in a dynamic spatial microsimulation model, the characteristics of the underlying population in the original dataset are aged so that factors like births, deaths and migration are modelled. This means that the policy change is made to a population that represents the 'best guess' population for the current and future times.

Chapter 2 of this book outlines how to prepare data for a spatial microsimulation model. This is an important step for any spatial microsimulation model. Without excellent data preparation, a model will be much harder to construct, and results will not be reliable. In any modelling, it is true to say that what comes out is only as good as what goes in.

Chapters 3–10 outline a number of methods for static spatial microsimulation, present a projection methodology for static spatial microsimulation, show how static spatial microsimulation models can be linked to aggregate or macro models and canvass the limits of static spatial microsimulation.

Chapters 11–14 show a number of methods that have been used for dynamic spatial microsimulation and then outline the limits of dynamic spatial microsimulation models.

Chapter 15 shows how spatial microsimulation models can be validated, and Chap. 16 provides an insightful background to the development of spatial microsimulation models and outlines the likely future directions for spatial microsimulation models.

1.2 History of Spatial Microsimulation

While some early modelling attempts could be considered to be spatial microsimulation (Hagerstrand 1952; Wilson and Pownall 1976), one of the first spatial microsimulation models was a model for health-care planning developed by Clarke et al. (1984). The model, called *HIPS (health information and planning system)*, was developed for the British health district authorities. The model generated an initial population from aggregate data for each location. The demographics of this initial population were then updated each year.

Clarke was also involved in other papers on spatial microsimulation modelling, including a spatial microsimulation model developed with Birkin called 'Synthesis' which used an iterative proportional fitting method, as described in Chap. 4 of this book (Clarke and Wilson 1985; Clarke and Holm 1987; Birkin and Clarke 1988; Birkin and Clarke 1989).

The next step in spatial microsimulation was a model developed by Clarke and Williamson, which was developed to estimate demand for water (Clarke et al. 1997;

Williamson et al. 1998). This model used a method developed by Williamson called combinatorial optimisation, which is described in Chap. 3 of this book.

Around this period, there was a considerable amount of work being done using spatial microsimulation. It was being used to look at regional changes in household incomes (Caldwell et al. 1998), for population projections (Van Imhoff and Post 1998) and for estimating household attributes (Williamson et al. 1998; Ballas and Clarke 1999; Ballas et al. 1999). All these models were static spatial microsimulation models and used either an iterative proportional fitting method, as described in Chap. 4, or a probabilistic combinatorial optimisation method, as described in Chap. 3. Other static spatial microsimulation models using combinatorial optimisation include the 'SMILE' model from Ireland (Ballas et al. 2005), as described in Chap. 7.

Another method being used for static spatial microsimulation is a deterministic reweighting method using a generalised regression method to reweight the survey weights provided on many survey files to small area benchmarks. This method has been pioneered by the National Centre for Social and Economic Modelling in Australia (Harding et al. 2003; Tanton 2007), and this method is described in Chap. 6. In terms of static spatial microsimulation, the final method described in this book is a deterministic combinatorial optimisation method (Edwards and Clarke 2009), and this is described in Chap. 5.

The first dynamic spatial microsimulation model was created by staff at the Spatial Modelling Centre in Sweden in 1999 (Vencatasawmy et al. 1999). This model is called 'SVERIGE' and is described in Chap. 12. In Britain, a number of dynamic spatial microsimulation methods have been developed, and these include 'Moses' (see Chap. 11) and 'MicroMaPPAS' (see Chap. 13). Also, 'SimBritain' has been developed by Ballas as a dynamic microsimulation model for Britain (Ballas et al. 2007). Note that many of these dynamic models originated as early static spatial microsimulation models and were subsequently developed into dynamic models.

There are limitations with both static and dynamic spatial microsimulation models, and no book is going to be complete without a discussion of these limitations, which are outlined in Chap.10 for static spatial microsimulation models and Chap. 14 for dynamic spatial microsimulation models. This discussion means that the reader is aware of what the limitations are before embarking on the construction of a model of this type.

1.3 Applications of Spatial Microsimulation Models

There are three main applications for spatial microsimulation models:

1. Small area estimation
2. Small area projection
3. Small area policy modelling

Deriving estimates for small areas is a key concern for governments at all levels, whether for tax, health, poverty alleviation or identifying areas of disadvantage. Most government services can be targeted to small areas, and information on where services are needed is imperative for efficient service delivery. Often these data do not exist at a small area level, and thus estimating these data is necessary. Small area estimation is a technique used by statisticians to derive estimates for one variable, but spatial microsimulation is now becoming a technique to allow the estimation of cross-tabulated data. So, while small area estimation can provide an estimate of incomes (e.g. Bell et al. 2007), spatial microsimulation can provide estimates of income cross-tabulated with, for example, family type (Miranti et al. 2011).

Projections of different populations are also vital for government planning, particularly as overcrowding becomes a major issue in some western countries and concerns around how to support an ageing population are raised. Dynamic spatial microsimulation can be used to look at where growth in populations that will require services will be in the future, so governments can plan for the future. Examples are looking at where older, single people will be living in the future, examining the need for aged care services (Lymer et al. 2009) or looking at where young children with all parents working will be living to assess demand for childcare places (Harding et al. 2011). Chapter 9 in this book shows how projections can be derived from static spatial microsimulation models.

Another application of spatial microsimulation is to look at where a policy will have the greatest effect. A recent example in Australia was looking at the reduction in poverty rates for older, single people on an age pension after an increase in the age pension (Tanton et al. 2009).

1.4 Validation of Spatial Microsimulation Models

An important part of any statistical model is validating the model against real-world data. There is not much point using a model if it does not represent the real world in some way. Validation is also very important for governments who use the output from microsimulation models, as they need to be sure that any modelled changes that they are basing their policies on are accurate and reliable. Chapter 15 surveys the different validation methodologies that have been implemented in different spatial microsimulation models.

1.5 The Future

There are two chapters that point to the future of spatial microsimulation modelling. One is the linking of static spatial microsimulation models to macro models (Chap. 8). This has been done with national microsimulation models (Bourguignon et al. 2010; Colombo 2010; Hérault 2010) but has not been done with spatial microsimulation models, so this is a fascinating new area for research.

Chapter 16 summarises what we see as the future for spatial microsimulation modelling. It is an exciting, and developing, field with strong applications in government and research. In essence, this is what drives many of the researchers in this field: the ability for spatial microsimulation models to inform government. These models really can make a difference in society by modelling potential policies and testing the spatial effect of these policies before they are implemented. This allows governments to see which areas are going to be affected the most by the proposed policy.

1.6 Conclusion

Spatial microsimulation modelling is an important development, enabled largely through increased computational capabilities. It provides valuable data to governments and researchers, allowing them to access detailed data at the small area level in order to understand the impact of policy changes and health outcomes in targeted populations.

References

Ballas, D., & Clarke, G. (1999, 23–27 August). *Regional versus local multipliers of economic change? A microsimulation approach*. 39th Regional Science Association (ERSA) Congress, University College, Dublin.

Ballas, D., Clarke, G., & Turton, I. (1999, 25–28 July). *Exploring microsimulation methodologies for the estimation of household attributes*. 4th International conference on GeoComputation, Mary Washington College, Fredericksburg, VA.

Ballas, D., Clarke, G., & Wiemers, E. (2005). Building a dynamic spatial microsimulation model for Ireland. *Population, Space and Place, 11*(3), 157–172.

Ballas, D., Clarke, G., Dorling, D., & Rossiter, D. (2007). Using SimBritain to model the geographical impact of national government policies. *Geographical Analysis, 39*(1), 44–77.

Bell, W., Basel, W., Cruse, C., Dalzell, L., Maples, J., O'Hara, B., & Powers, D. (2007). *Use of ACS data to produce SAIPE model-based estimates of poverty for counties*. Washington, DC: U.S. Census Bureau.

Birkin, M., & Clarke, M. (1988). SYNTHESIS – A synthetic spatial information system for urban and regional analysis: Methods and examples. *Environment and Planning A, 20*(12), 1645–1671.

Birkin, M., & Clarke, M. (1989). The generation of individual and household incomes at the small area level using synthesis. *Regional Studies: The Journal of the Regional Studies Association, 23*(6), 535–548.

Bourguignon, F., Bussolo, M., & Cockburn, J. (2010). Guest Editorial: Macro–micro analytics: Background, motivation, advantages and remaining challenges. *International Journal of Microsimulation, 3*(1), 1–7.

Caldwell, S. B., Clarke, G. P., & Keister, L. A. (1998). Modelling regional changes in US household income and wealth: A research agenda. *Environment and Planning C: Government and Policy, 16*(6), 707–722.

Clarke, M., & Holm, E. (1987). Microsimulation methods in spatial analysis and planning. *Geografiska Annaler Series B, 69*(2), 145–164.

Clarke, M., & Wilson, A. (1985). The dynamics of urban spatial structure: The progress of a research programme. *Transactions of the Institute of British Geographers, 10*(4), 427–451.

Clarke, M., Forte, P., Spowage, M., & Wilson, A. G. (1984). A strategic planning simulation model of a district health service system: The in-patient component and results. In W. van Elmeren, R. Engelbrecht, & C. D. Flagle (Eds.), *Systems science in health care*. Berlin: Springer.

Clarke, G. P., Kashti, A., McDonald, A., & Williamson, P. (1997). Estimating small area demand for water: A new methodology. *Water and Environment Journal, 11*(3), 186–192.

Colombo, G. (2010). Linking CGE and microsimulation models: A comparison of different approaches. *International Journal of Microsimulation, 3*(1), 72–91.

Edwards, K. L., & Clarke, G. P. (2009). The design and validation of a spatial microsimulation model of obesogenic environments in Leeds: SimObesity. *Social Science and Medicine, 69*, 1127–1134.

Hagerstrand, T. (1952). The propagation of innovation waves. http://catalogue.nla.gov.au/Record/4364370. Accessed 15 Aug 2011.

Harding, A., Lloyd, R., Bill, A., & King, A. (2003, 30–31 May). *Assessing poverty and inequality at a detailed regional level: New advances in spatial microsimulation*. Inequality, Poverty and Human Well-Being Conference, Helsinki, Finland.

Harding, A., Vidyattama, Y., & Tanton, R. (2011). Demographic change and the needs-based planning of government services: Projecting small area populations using spatial microsimulation. *The Journal of Population Research, 28*(2–3), 203–224.

Hérault, N. (2010). Sequential linking of computable general equilibrium microsimulation models: A comparison of behavioural reweighting techniques. *International Journal of Microsimulation, 3*(1), 35–42.

Lymer, S., Brown, L., Harding, A., & Yap, M. (2009). Predicting the need for aged care services at the small area level: The CAREMOD spatial microsimulation model. *International Journal of Microsimulation, 2*(2), 27–42.

Miranti, R., McNamara, J., Tanton, R., & Harding, A. (2011). Poverty at the local level: National and small area poverty estimates by family type for Australia in 2006. *Applied Spatial Analysis and Policy, 4*(3), 145–171.

Orcutt, G., & Glazer, A. (1980). Microanalytic modelling and simulation. In B. Bergmann, G. Eliasson, & G. Orcutt (Eds.), *Simulation models: Methods and applications*. Stockholm: Industrial Institute for Economic and Social Research.

Orcutt, G., Greenberger, M., Korbel, J., & Rivlin, A. (1961). *Microanalysis of socioeconomic systems: A simulation study*. New York: Harper and Row (Reproduced in International Journal of Microsimulation, 2007, *1*(1), 3–9).

Sutherland, H. (1995). *Static microsimulation models in Europe*. Cambridge Working papers in Economics 9523. Faculty of Economics, University of Cambridge, Cambridge.

Tanton, R. (2007, August). *SPATIALMSM: The Australian spatial microsimulation model*. Paper presented at the 2nd International Microsimulation Conference, Vienna, Austria.

Tanton, R., Vidyattama, Y., McNamara, J., Vu, Q., & Harding, A. (2009). Old, single and poor: Using microsimulation and microdata to analyse poverty and the impact of policy change among older Australians. *Economic Papers: A Journal of Applied Economics and Policy, 28*(2), 102–120.

Van Imhoff, E., & Post, W. (1998). Microsimulation methods for population projection. *Population: An English Selection, 10*(1), 97–138.

Vencatasawmy, C. P., Holm, E., Rephann, T., Esko, J., Swan, N., Öhman, M., Åström, M., Alfredsson, E., Holme, K., & Siikavaara, J. (1999, September 3–7). *Building a spatial microsimulation model*. Paper presented at the 11th European Colloquium on Quantitative and Theoretical Geography, Durham, England.

Williamson, P., Birkin, M., & Rees, P. (1998). The estimation of population microdata by using data from small area statistics and samples of anonymised records. *Environment and Planning A, 30*(5), 785–816.

Wilson, A., & Pownall, C. (1976). A new representation of the urban system for modelling and for the study of micro-level interdependence. *Area, 8*, 246–254.

Zaidi, A., Harding, A., & Williamson, P. (2009). *New frontiers in microsimulation modelling*. Vienna: Ashgate.

Chapter 2
Building a Static Spatial Microsimulation Model: Data Preparation

Rebecca Cassells, Riyana Miranti, and Ann Harding

This chapter provides practical instruction and examples of some of the key issues that need to be considered when selecting and preparing data for building a spatial microsimulation model. The data underlying the spatial microsimulation model being built are fundamental to the accuracy and robustness of the small area results produced from the model. If these data are correct and as compatible as possible, the procedures being used to produce the spatial microsimulation model will operate more effectively, and the output will be more reliable.

There are numerous issues that may develop around data preparation for the model; some will have simple solutions, while others will require more thought and resources. Some of the considerations and issues that may arise include data requirements and data sources, resolving differences in variable definitions and data scope and the absence of a particular population in the dataset being used (e.g. children).

As discussed in Chap. 1, spatial microsimulation models are typically constructed by using data from a nationally representative sample survey and some reliable small geographic area data source, normally a national Census. There are variations around this; for example, some models use synthetically created data rather than sample survey data. While every country will have different national sample surveys and varying national population Censuses, the following is a checklist of issues that have emerged during NATSEM's (National Centre for Social and Economic Modelling) Australian attempts to prepare various national sample surveys for reweighting to the Australian Census tables. Many of the examples provided in this chapter relate to NATSEM's spatial microsimulation model, SpatialMSM, but can be equally relevant to other spatial microsimulation models.

R. Cassells (✉) • R. Miranti • A. Harding
National Centre for Social and Economic Modelling (NATSEM), University of Canberra, Canberra, Australia
e-mail: Rebecca.cassells@natsem.canberra.edu.au

R. Tanton and K.L. Edwards (eds.), *Spatial Microsimulation: A Reference Guide for Users*, Understanding Population Trends and Processes 6, DOI 10.1007/978-94-007-4623-7_2,

2.1 Data Sources and Requirements

Two data sources are typically required to build a spatial microsimulation model. The first is a representative sample survey, which provides a wide variety of rich and detailed information but lacks geographic information—such as a survey of income and housing. The second dataset is one which often has limited data items and detail, but high geographic disaggregation—such as a Census. Typically both datasets will be nationally representative; however, spatial microsimulation models for specific areas within a country or region may also be built if adequate data is available.[1] From this point on, we will refer to the two common datasets used in spatial microsimulation as a 'sample survey' and a 'Census'.

Data from the geographically rich dataset are usually used as benchmarks or constraints, to which synthetic small geographic area estimates produced by the spatial microsimulation process must match.

Datasets may also be combined in order to maximise the sample size available for modelling and improve the output from the microsimulation model. For example, the latest versions of the Australian spatial microsimulation model—SpatialMSM—combined two surveys of income and housing. This gives greater reliability to the model but also means the base data for the SpatialMSM model is compatible with the base data for NATSEM's static microsimulation model—STINMOD. This model replicates the rules of the Australian income tax, social security and family payment programmes (Lloyd 2007; Percival et al. 2007). By doing this, the model then has the added potential of providing policy analysis at a small geographic area level, if the research question calls for this (Harding et al. 2009; Tanton et al. 2009).

Additional data sources may also be necessary in order to facilitate special population imputation (see Sect. 2.4 below).

2.2 Sample Scope

One of the most important things in a spatial microsimulation method is ensuring that the scope of the survey sample and the Census are the same or can be amended to be the same. There are several common areas where the two datasets may not have the same scope. These include:

1. The unit of analysis (household, income unit, etc.).
2. The inclusion or exclusion of persons in non-private dwellings, like hospitals or nursing homes, which are usually on the Census information but are not usually in survey samples.

[1] A nationally representative survey is one that has been designed and conducted in a way so that it captures the characteristics of all persons and households. Households and persons within the survey are then assigned a 'weight', which, when summed, will equal the entire population of that nation.

3. The classification of households—for example, in the Australian Census, there are 'non-classifiable' households, which are households that satisfy one of a number of criteria which means the Census data may be unreliable. These criteria include having no one in the house aged over 15 and having inadequate data on the Census form.

Depending upon the data sources being used, the scope differences will vary, and it is important that these differences are identified and corrected as much as possible. These areas are discussed further below.

2.3 Unit of Analysis

In many cases, the sample survey and the Census may use the same income-sharing unit (the unit within which income is assumed to be shared). However, this is not always the case. It needs to be confirmed that a 'household' in the sample survey has the same meaning as a 'household' in the Census. In the Australian case, NATSEM's STINMOD model uses a special 'social security'-type income unit, which is a subset of the usual household income-sharing unit. In addition, some of the earlier ABS sample surveys used a nuclear family income unit definition, which was again a subset of the standard household income-sharing unit. In all such cases, the smaller income units had to be aggregated to household units before the spatial microsimulation model could be run.

2.3.1 Non-private Dwellings

While not universally the case, most national sample surveys only include the population that live in private dwellings within their scope. Thus, for example, the Australian Bureau of Statistics' (ABS) Survey of Income and Housing (SIH) samples households and individuals resident in private dwellings and excludes those resident in non-private dwellings such as aged care and nursing homes, prisons, boarding schools and hospitals. The ABS Census includes people living in non-private dwellings. If the sample survey file of households residing in occupied private dwellings is used in combination with information from the Australian Census, then the spatial microsimulation model will be biased.

For example, suppose that we are trying to derive estimates of health status by age for each small geographic area by using a spatial microsimulation model. If a particular small geographic area contains five large nursing homes and we use 'age by gender' Census totals as a benchmark for that small geographic area, then we are likely to overstate the number of healthy over-70-year-olds actually residing in that small area. This is because the relatively healthy over-70-year-olds are likely to be still living in their own homes, which means the sample survey will have been inflated to match the total number of over-70-year-olds as shown in the Census results for that small area (a total which will include the relatively unhealthy over-70-year-olds in nursing homes). Thus, it is important to ensure consistency between the population in scope in the relevant sample survey and the population in scope for the Census.

In order to achieve this consistency, it may be necessary to remove or impute information about particular populations. For some analyses, information about people in non-private dwellings may be required. For example, the area of interest may be people aged over 80 years; however, a substantial proportion of these people often reside in hospitals or nursing homes. As discussed above, often people living in non-private dwellings are excluded from survey data but are included in Census data. In order to 'match' these populations, it will be necessary to develop a methodology for removing or including these persons, depending upon the research question or operation required by the model. More information on this process is provided in Sect. 2.4.2 below.

2.3.2 Non-classifiable Households

Another area where the scope of the sample survey and Census may not match is when people are grouped as living in 'other non-classifiable households' or some such similar classification. In Australia, these households are defined as those households that contain no persons aged over 15 years; that the collector deemed occupied but was unable to make contact with any occupants; or where the information supplied on the Census form was inadequate. This discrepancy between the two data sources was resolved by obtaining special benchmark tables from the ABS that excluded non-classifiable households. Other methods may need to be employed in the case that these data are not available. In previous microsimulation models constructed at NATSEM, a non-classifiable population was created for the survey; however, this was considered to be an inferior solution.

2.4 Population Imputation

As discussed above, population imputation may be necessary for a spatial microsimulation model if there are specific populations of interest missing from either dataset. The most common populations of interest that are often absent from sample surveys but included in Censuses are children and persons living in non-private dwellings. Imputation of these two important populations is discussed below.

2.4.1 Imputation of Child Records

Typically, in ABS surveys, only persons aged 15 and over are included in sample surveys, and no individual records of children are available. The records for children within a household are held with the household information, and limited information on children is available. Imputing child records onto the survey will enable child-focused research to take place, such as estimating the numbers of children in poverty for particular communities.

The problem of not having children on the survey can be overcome somewhat by imputing the number of children in each household through existing household information from the survey which shows how many children in each age group reside in each household. Where this information is not available, alternative imputation methods can be used (see Brick and Kalton 1996 for a description of methods); however, more often than not, the record for the household will have information about the number of and ages of children residing in each household.

These known dependent children can then be output as individual person-level records to the sample survey and assigned to their corresponding households. Each child record will need to be assigned a new individual identifier; however, the family identifier, income unit identifier, household identifier and the age group variables (where they exist) can be retained. Relevant household and income unit variables can then be merged onto the child-level dataset and appropriate values for person-level variables assigned to each child. For example, the occupation and income fields (if included as benchmarks) will be assigned a value of zero (for not in labour force) and current study status a value that reflects full-time student for those of school age. Some values may need to be randomly assigned based on known values or ratios—such as sex.

Another issue that may be encountered when imputing child records is the top-coding of variables, which is often carried out by national statistical agencies in an effort to maintain data confidentiality. For example, in the 2007–2008 Survey of Income and Housing, the ABS top-coded the number of children in each age range, resulting in a capped total number of children in the household of five or more children. To overcome this, the total number of usual residents in each household, together with the total number of children and adults, is used to re-estimate the number of children. Where it has been determined that a household had an 'extra' child, this child was randomly assigned within the child age ranges available. This imputation method retains the ABS confidentiality while also providing the information required for the model.

2.4.2 Imputation of a Non-private Dwelling Population

People living in non-private dwellings (NPDs) are often an important population to include in an analysis. These people, while being a relatively small population compared to people in occupied private dwellings (OPD), are often recipients of income support and are therefore of interest to researchers and policymakers. Survey data does not often have information about persons living in NPDs (which includes dwellings such as hospitals, boarding schools, prisons and nursing homes), whereas Census data will typically include this information. Given this inconsistency, information about non-private dwellings can either be deleted from or added to each data source in order to make them directly comparable.

If there is a need to create a synthetic non-private dwelling population, this can be done using existing data about persons living in NPDs. For the SpatialMSM model, special records were created for individuals in non-private dwellings by using information available in the 2001 Census 1% unit record file (Cassells et al. 2010). These records were then attached to the survey unit record file. Both children and adults residing in NPDs were included in the sample, and only those persons classified as usual residents were included in the NPD population. Detail about these persons was imputed from other known values—for example, a single integer value of income was imputed from the income range available in the sample file. Most other person-level detail was available, for example, labour force status, age, number of hours worked, study status, type of educational institution attending and so on. Persons in NPDs obviously receive a value of zero for all household and family-level variables. Each NPD record was assigned a weight of 100, given that the data has been derived from a 1% random sample of the 2001 Census. This resulted in 1,995 adult (persons aged 15 and over) NPD records and 109 child (persons aged under 15 years) NPD records.

2.5 Matching Variable Definitions in the Sample Survey and the Census

For the spatial microsimulation process to work correctly, the variables used to match in both the Census and the survey dataset must be defined in the same way.

One key issue here involves matching variable definitions used in the sample survey and the Census small area tables. In some cases, this may require aggregating finer groups contained in either the Census or the sample survey to broader aggregations. For example, the sample survey may have eight categories of post-school qualifications, while the Census may have only four. Careful reading of the documentation for both data sources is required to correctly aggregate the various categories so that they match exactly—a process which may, for example, ultimately end up with only two post-school qualification categories in both data sources.

A second issue is that variables that at first glance appear the same—for example, 'labour force status'—may be defined quite differently in the two data sources. For example, one data source may consider being 'unemployed' as working no hours of paid work per week, and another data source may define 'unemployed' as those who receive unemployment benefits from the government. For SpatialMSM, because both data sources are from the same statistical agency (the Australian Bureau of Statistics), variables are often defined in the same way, as ABS standard definitions are applied across all surveys and the Census. However, this may be a more significant issue if the data were from different agencies.

A third issue will arise if a dataset has non-response values and another does not. These non-response values are often referred to as 'not-stated' values. For example, due to the nature of collection of Australian Census data (non-interviewer-assisted),

Census data often contains fully and partially not-stated values. However, any partial non-response value from ABS survey data is imputed. This being the case, the non-response values from the Census need to be redistributed amongst known categories so that the two data sources can be as compatible as possible. For SpatialMSM, this redistribution was proportional, based on the relative frequency of known values. Other methods may also be employed to assign not-stated values appropriately.

2.6 Uprating and Deflating

Uprating or deflating typically involves adjusting monetary values collected within the sample survey to account for estimated price movements since the time of the survey or anticipated future movements (Harding 1996, p. 3). For example, a static microsimulation model that is trying to capture the 2011–2012 tax and transfer systems may be built upon 2009 sample survey data. Here, the earnings of employees shown in the survey data need to be inflated by movements in average weekly earnings between 2009 and 2011–2012. Alternatively, the data may require deflating, if the sample survey is more current than the benchmark data. All dollar values used in the benchmark tables will need to be adjusted by a suitable value. For SpatialMSM, housing costs (rent and mortgages) and personal and household incomes are adjusted using consumer price index and average weekly earnings changes, respectively.

To be able to match the sample survey data to Census data, all the financial data have to relate to the same year. In most cases, it is easier to uprate the financial data in the surveys to the year of the Census rather than adjust the Census data.

2.7 Balancing Data

National statistical agencies will often randomise small area data to maintain a level of confidentiality. This randomisation will often result in slightly different population totals for the benchmark tables being used. In NATSEM's previous spatial microsimulation models, a complicated and time-consuming process took place in order to align the Census benchmark total populations, as this was thought to improve the reweighting process and convergence. This process was termed 'balancing' the tables. In 2007, some sensitivity analysis was conducted in order to determine if there was a significant bias in the results produced through using unbalanced data. A comparison of balanced and unbalanced results from the Australian Capital Territory (ACT) in Australia found that the practice of balancing data had little effect on the results and in some cases, the results obtained from the unbalanced data were closer to the true Census counts than those from the balanced data. Given the results of the sensitivity analysis, the current Census benchmark tables used for SpatialMSM are not balanced. However, work on projecting data using a spatial microsimulation model, reported in Chap. 9 of this book, has found that balancing becomes more important for data projections, particularly for longer-term projections, and balancing is used in this instance.

2.8 Conclusion

This chapter details issues and specific measures taken to prepare and harmonise sample survey and Census data needed to build a spatial microsimulation model.

Transforming and manipulating these data sources, so that they are as compatible as possible, will ensure that the spatial microsimulation technique being used is optimised, and the output gained from the model will be as accurate as possible. The issues discussed in this chapter are those experienced with a particular model using Australian data and are not exhaustive. Other issues may arise depending upon each model being built and the raw data being used.

References

Brick, J. M., & Kalton, G. (1996). Handling missing data in survey research. *Statistical methods in medical research, 5*, 215–238. Sage publications, United Kingdom.

Cassells, R., Harding, A., Miranti, R., Tanton, R., & McNamara, J. (2010). *Spatial microsimulation: Preparation of sample survey and census data for SpatialMSM/08 and SpatialMSM/09* (NATSEM Technical Paper 36). https://guard.canberra.edu.au/natsem/index.php?mode=download&file_id=1065

Harding, A. (Ed.). (1996). *Microsimulation and public policy* (Contributions to economic analysis series). Amsterdam: North Holland.

Harding, A., Vu, N. Q., Payne, A., & Percival, R. (2009). Trends in effective marginal tax rates in Australia from 1996–97 to 2006–07. *Economic Record, 85*(271), 449–461.

Lloyd, R. (2007). STINMOD: Use of a static microsimulation model in the policy process in Australia. In A. Harding & A. Gupta (Eds.), *Modelling our future: Population ageing, social security and taxation* (International symposia in economic theory and econometrics, Vol. 16). Amsterdam: North Holland.

Percival, R., Abello, A., & Vu, Q. (2007). STINMOD (Static income model). In A. Gupta & A. Harding (Eds.), *Modelling our future: Population ageing, health and aged care* (International symposia in economic theory and econometrics, Vol. 16, pp. 477–482). Amsterdam: North Holland.

Tanton, R., Vidyattama, Y., McNamara, J., Vu, Q. N., & Harding, A. (2009). Old, single and poor: Using microsimulation and microdata to analyse poverty and the impact of policy change among older Australians. *Economic Papers: A Journal of Applied Economics and Policy, 28*(2), 102–120.

Part II
Static Spatial Microsimulation Models

Chapter 3
An Evaluation of Two Synthetic Small-Area Microdata Simulation Methodologies: Synthetic Reconstruction and Combinatorial Optimisation

Paul Williamson

3.1 Background

Population microdata comprise a list of individuals, normally nested into families and households, with each individual having an associated set of personal demographic and socio-economic characteristics. Microdata offer well-known advantages over tabular data, including enhanced possibilities for linkage to other data sources, retention of maximum flexibility in user-determined aggregation and analysis and efficiencies in data storage (for large multivariate datasets) (Birkin and Clarke 1995). These advantages are reflected in the widespread use by researchers of public use microdata from censuses and social surveys.

To protect respondent confidentiality, population microdata are typically stripped of all subregional geography. However, a clear demand exists for microdata spatially coded to subregion level. This demand is reflected in calls for a third, more spatially detailed, Sample of Anonymised Records from the UK 2001 Census (Dale and Teague 2002). In the absence of such data, a number of projects have been forced to generate their own synthetic small-area population microdata (Birkin and Clarke 1988; Beckman et al. 1996; Martin et al. 2001).

Four main approaches to the creation of spatially detailed synthetic population microdata may be identified: data fusion/merging, stratified sampling, reweighting and imputation (Williamson 2002). Of these four approaches, two are not practicable. Data fusion and merging requires levels of access to the original microdata not normally permissible due to legal safeguards on respondent confidentiality, whilst shortcomings in published small-area data mean that conventional stratified sampling is unable to capture the highly complex and multidimensional nature of between-area differences (Voas and Williamson 2001a). The statistical reliability of the two

P. Williamson (✉)
Department of Geography, School of Environmental Sciences,
University of Liverpool, Liverpool, UK
e-mail: P.Williamson@liv.ac.uk

R. Tanton and K.L. Edwards (eds.), *Spatial Microsimulation: A Reference Guide for Users*, Understanding Population Trends and Processes 6, DOI 10.1007/978-94-007-4623-7_3,

remaining approaches, although both in widespread use, has never been systematically and rigorously evaluated.

Section 3.2 introduces the two approaches to the creation of synthetic small-area population microdata evaluated in this chapter. The first, synthetic reconstruction, is an imputation-based approach. The second, combinatorial optimisation, is a reweighting-based approach. Fuller accounts of the original implementations of each approach may be found elsewhere (Birkin and Clarke 1988; Williamson et al. 1998). As part of the research reported in this chapter, a number of significant technical improvements have been made to these methods, all aimed at maximising their performance. These innovations are detailed in Sects. 3.3 and 3.4. Both methods share in common the use of known local area information as 'constraints' on their estimates, in an attempt to capture between-place variations in population characteristics. Therefore, Sect. 3.5 reviews the nature of between-area diversity and within-area homogeneity, to help better understand the nature and number of local area constraints required. This is followed, in Sect. 3.6, by the presentation of a new framework for the evaluation and validation of small-area synthetic microdata, which includes innovations in both the measures of fit used, and in the types of fit measured. Sections 3.7 and 3.8 present an evaluation of the relative performance of synthetic reconstruction and combinatorial optimisation, before the chapter concludes (Sect. 3.9) with some general comments on strengths, weaknesses and potential utility, of the resulting synthetic small-area microdata.

3.2 Synthetic Reconstruction and Combinatorial Optimisation Methodologies

3.2.1 Synthetic Reconstruction

Synthetic reconstruction (SR) is an imputation-based approach. For the small area of interest, one census tabulation is used to provide an initial list of individuals with a set of known population attributes (e.g. the age and sex of each population member). All other attributes are added (imputed) by sampling from probabilities conditional upon one or more previously generated attributes. In as far as is possible, these conditional probabilities are derived from published small-area census tabulations but where necessary draw upon tabulations published for higher-level geographies.

Figure 3.1 illustrates the basic process in more detail. In step 1, the census count of the number of persons, by age, sex and marital status, is found for the small area being synthetically estimated. In the example given, this count included one married male aged 18. Hence, a first person is created and assigned these attribute values (highlighted in bold in Fig. 3.1). In steps 2–4, the employment status of this person is imputed. In step 2, the probability of the person having each possible employment status is identified. In this illustrative example, the probabilities are conditional upon age group, sex and marital status. Hence, they are different for each of the three persons shown. These probabilities are then converted into a set of cumulative

Steps	Person: 1st	Person: 2nd	Person: Last
1. Age, sex and marital status of person[a]	Age: **18** Sex: **Male** M: **Married**	Age: 34 Sex: Male M: Married	Age: 87 Sex: Male M: Married
2. Probability of employment status given age, sex and marital status *E: employed* *U: unemployed* *I: inactive*	*Prob* / *Cum Prob Bin* E: 0.4 (0.0-0.4) U: 0.3 (>0.4-0.7) I: 0.3 (>0.7-1.0)	*Prob* / *Cum Prob Bin* E: 0.6 (0.0-0.6) U: 0.3 (>0.6-0.9) I: 0.1 (>0.9-1.0)	*Prob* / *Cum Prob Bin* E: 0.0 (0.0-0.0) U: 0.0 (0.0-0.0) I: 1.0 (>0.0-1.0)
3. Random number *(computer generated)*	0.281	0.709	0.481
4. Employment status assigned on basis of random sampling	Employed	Unemployed	Inactive
5. Next person *(repeat until all persons have been assigned an employment status)*	*Move on to next person*	*Move on to next person*	*End*

Fig. 3.1 The synthetic reconstruction approach (Adapted from Clarke 1996)

probability 'bins'. In steps 3 and 4, a random draw is made, such that whichever cumulative probability 'bin' the random number falls within determines the person's imputed employment status. For the first person, the random number (0.281) falls within the first cumulative probability 'bin' (0.0–0.4). Hence, he is imputed the employment status of 'employed'. The synthetic reconstruction process then moves on to create additional persons, in similar fashion, until the known local area count of persons, by age, sex and marital status, has been satisfied. If additional attributes are required, these are then imputed in similar fashion to employment status. The main challenges associated with this approach include (1) producing the known or estimated 'local area' conditional probabilities required to impute additional attributes and (2) placing persons within families and households (if required).

3.2.2 Combinatorial Optimisation

The second approach, combinatorial optimisation (CO), involves the selection of a combination of households from an existing survey micro data set that best fit published small-area census tabulations. In effect, this is an integer reweighting approach, in which most households are assigned weights of zero (i.e. not present). The process involves a number of steps, as outlined below.

Step 1 *Obtain sample survey microdata*
Combinatorial optimisation can take as in input any survey microdata, such as a national government survey, provided that the microdata contain (1) attributes of interest for post-estimation analyses and (2) attributes that map onto the small-area constraints to be used in the estimation process (c.f. Step 2). A simplified example of such survey microdata is given below, including only household-level records.

	Size	*Adults*	*Children*
Household A	2	2	0
Household B	2	1	1
Household C	4	2	2
Household D	1	1	0
Household E	3	2	1

More commonly, the survey also includes person-level records, nested within families and households. In this latter case, selection of a household automatically includes selection of all of the family and person records within that household.

Step 2 *Identify small-area constraints*
A selected set of known small-area counts are used to act as constraints on the estimation process. These *observed* counts are typically a subset of published census counts/tabulations for the small area being estimated. All of the constraint tables used should relate to the same small area, although inconsistencies in counts between tables – arising, for example, as a result of prepublication disclosure control measures – are permissible.

Constraint Table 1: household size (persons per household)		Constraint Table 2: age of occupants	
Household size	Frequency	Type of person	Frequency
1	1	Adult	3
2	0	Child	2
3	0		
4	1		
5+	0		
Total	**2**		

Step 3 *Randomly select households from the sample survey*
The synthetic small-area microdata generated by combinatorial optimisation comprise a selected set of households from the sample survey. The precise number of households to be selected is defined by one of the chosen constraint tables. In this case, because constraint Table 3.1 (household size) records the small area being estimated as comprising *two* households, *two* households are randomly selected:

	Size	*Adults*	*Children*
Household A	2	2	0
Household E	3	2	1

Table 3.1 Constraints used during synthetic microdata generation

Table constraints used in final version of Pop91CO		Tables used for comparison of Pop91SR and Pop91CO	
Census tables	Variables in tabulation	As inputs	For assessment of fit
S35	Age/sex/marital status	•	•
S42	Household composition/tenure	•	•
S86	Socio-economic group of head/tenure	•	•
S06	Age/ethnic group	•	X
S08	Age/sex/economic position	•	•
S09	Sex/economic position/ethnic group	•	X
S34	Sex/marital status/economic position	•	•
S39	Age/sex/marital status of household head	•	•
S49	Ethnic group of household head/tenure	•	•

Step 4 *Convert the synthetic microdata into estimated counts*
In order to assess the fit of the synthetic microdata to known small-area constraints, it is first necessary to tabulate the synthetic small-area microdata to find the *estimated* equivalents of the counts observed in the small-area constraint tables:

Estimated household size (persons per household)		*Estimated* age of occupants	
Household size	Frequency	Type of person	Frequency
1	0	Adult	4
2	1	Child	1
3	1		
4	0		
5+	0		
Total	**2**		

Step 5 *Assess the fit of the synthetic microdata to constraint data*
Armed with the estimated counts (step 4) and observed counts (step 2) for each constraint table, it is now possible to calculate the size of the difference between them and to sum these tabular differences across all constraint tables to find the total absolute difference.

Constraint 1: household size	Estimated frequency (i)	Observed frequency (ii)	Absolute difference |(i)–(ii)|
1	0	1	1
2	1	0	1
3	1	0	1
4	0	1	1
5+	0	0	0
		Subtotal:	*4*

Constraint 2: age	Estimated frequency (i)	Observed frequency (ii)	Absolute difference |(i)–(ii)|
Adult	4	3	1
Child	1	2	1
		Subtotal:	*2*

Total absolute difference = *4* + *2* = **6**

Step 6 *Randomly swap one of selected households*
Next, in an attempt to improve the fit of the small-area synthetic microdata with the known small-area constraints, one of the currently selected survey households is randomly swapped with another household selected at random from the survey sample. In the example below, Household A has been replaced at random by Household D.

	Size	*Adults*	*Children*
Household D	1	1	0
Household E	3	2	1

This random household swapping can be conducted with or without replacement. Swapping with replacement means that the same survey household can be selected to appear more than once in the household combination constituting the synthetic microdata, which is equivalent to giving the household an integer weight of more than one. Swapping with replacement is recommended for reasons outlined in Sect. 3.4.5.

Step 7 *Assess the fit of the post-swap synthetic microdata*
The fit of this new household selection is assessed by once again comparing the observed small-area constrain to their synthetically estimated counterparts (i.e. by repeating steps 4–6 above).

Household size	Estimated frequency (i)	Observed frequency (ii)	Absolute difference |(i)–(ii)|
1	1	1	0
2	0	0	0
3	1	0	1
4	0	1	1
5+	0	0	0
		Subtotal:	*2*

Age	Estimated frequency (i)	Observed frequency (ii)	Absolute difference |(i)–(ii)|
Adult	3	3	0
Child	1	2	1
		Subtotal:	*1*

Total absolute difference = *2* + *1* = **3**

Step 8 *Accept or reject swap*
If the household swap leads to a worsening of the fit to local area constraints (i.e. an increased total absolute difference), then reverse the swap (i.e. remove household D and reinsert household A). Otherwise, retain the new household selection (D and E) as the current 'best estimate'.

Step 9 *Keep swapping households for as long as required*
In order to further improve the fit of the synthetic microdata, repeat steps 6–8 until no further reduction in total absolute difference is possible or at least until an acceptable level of fit between synthetic microdata and the chosen small-area constraints has been achieved.

Final selected households

	Size	*Adults*	*Children*
Household C	4	2	2
Household D	1	1	0

Final fit to known small-area constraints

Household size	Estimated frequency (i)	Observed frequency (ii)	Absolute difference |(i)–(ii)|
1	1	1	0
2	0	0	0
3	0	0	0
4	1	1	0
5+	0	0	0
		Subtotal:	*0*

Age	Estimated frequency (i)	Observed frequency (ii)	Absolute difference |(i)–(ii)|
Adult	3	3	0
Child	2	2	0
		Subtotal:	*0*

Total absolute difference = *0* + *0* = **0**

In outlining the basic approach to combinatorial optimisation, one key refinement has been side-stepped. The approach outlined above adopts a 'hill-climbing' algorithm, in which a household swap is only accepted if it improves the overall fit of the synthetic microdata to local constraints (c.f. Step 8). In reality, nearly all users of combinatorial optimisation prefer to adopt either a 'simulated annealing' or 'genetic' algorithm, in which swaps which adversely affect the fit *might be* accepted in order to avoid getting trapped with a suboptimal selection of households (Williamson et al. 1998).

3.3 Innovations in Synthetic Reconstruction

Section 3.2 outlined the basic approach to synthetic reconstruction. The performance of synthetic reconstruction can be optimised if a number of additional technical innovations are adopted. The innovations implemented for the research reported in this chapter are reported below.

3.3.1 Modified Monte Carlo Sampling

In conventional Monte Carlo sampling, potentially significant error in final outputs is introduced due to the stochastic nature of the sampling process. For example, even though a variable, x, might have five categories with a known proportional distribution, P, of {0.12, 0.25, 0.52, 0.04, 0.07}, one possible outcome of imputing this attribute to 20 (N) individuals is a synthetic distribution of {0, 0, 0, 0, 20}. A modified Monte Carlo sampling strategy has been devised which leads to an average 40% reduction in associated variance (Huang and Williamson 2001a). First, a target distribution is identified, which is equal to P times N. Thus, in the above example where N=20, the target distribution would be {2.4, 5, 10.4, 0.8, 1.4}. Of this target, the integer part is {2, 5, 10, 0, 1}, comprising a total count of 18. Given this integer target distribution, 18 of the 20 individuals awaiting imputation of a value for x are selected (in random order). The first two individuals selected are assigned to the first category of variable x; the next five to the second category of x and so on, such that distribution of imputed values {2, 5, 10, 0, 1} exactly equals the integer part of the target distribution. This leaves two individuals awaiting imputation and an unassigned target distribution of {0.4, 0, 0.4, 0.8, 0.4} (i.e. the factional part of the original target distribution). This fractional target is converted into a probability distribution, against which the value of variable x is imputed, in the usual manner, by random draw. The stochastic nature of this remaining phase of imputation is further reduced by reducing to zero the fractional target for any value of x once it has been randomly selected for imputation and recalculating the associated probability distribution before the next value of x is imputed. This has the effect of limiting the number of times a given value of x can be imputed during the 'fractional phase' of the imputation process to 1.

3.3.2 Statistical Justification of Reconstruction Order

Previous approaches to synthetic reconstruction have acknowledged the subjective way in which linkages between census tabulations and the order of data imputation have been selected. For this project, a mixture of logistic regression and CHAID analysis was used to identify the key determinants of target attributes and therefore to identify the most appropriate census tabulations for use in their reconstruction.

3.3.3 Modelled 100% Counts of 10% Data

Only a 10% sample of write-in questions, such as occupation, were coded for UK censuses prior to 2001. In previous synthetic reconstructions, relevant probabilities have been derived using these ED-level small-area data unmodified. Such an approach has been found to lead to severely biased results (Voas and Williamson 2000a). To address this problem, more extensive use has been made here of other known information: (1) the equivalent 10% tables from all of other EDs in the same ward, (2) the ward-level version of the 10% table of interest, and (3) any marginals for these 10% tables that have also published using the full 100% sample (e.g. age). First, a quasi 100% ward-level table is created by reweighting the counts in the 10% ward-level table to fit the sum of the known 100% ED-level margins. Iterative proportional fitting is then used to adjust the counts within the ED-level 10% tables, such that they agree with the counts in the quasi 100% ward-level table. These adjusted ED-level tables provide best estimates of the actual small-area distributions.

3.3.4 Improved Data Linkage

Small-area tabulations do not of themselves contain sufficient information to allow plausible synthetic reconstruction. For example, at ED level, it may not be possible to establish a direct link between two key population attributes. Instead, the missing information has routinely been drawn from tables published for higher geographical levels (ward, district, national) and combined with the available small-area data using iterative proportional fitting. For this chapter, a similar approach has been adopted. However, whereas previously data and time constraints have restricted researchers to combining information typically drawn from only two geographical levels and to creating conditional probabilities linking involving only three or four attributes at a time, for this chapter, the availability of Samples of Anonymised Records (SARs) from the 1991 UK Census has allowed information from three levels of geography (ED, ward and nation) to be combined to estimate conditional probabilities linking up to five attributes simultaneously – for example, ethnic group of household head conditional upon household head's age, sex, marital status, economic position and tenure. This greater linkage has reduced dependence on the assumption of conditional independence between related variables.

3.3.5 Data Reconciliation

To protect respondent confidentiality, all small-area census counts are subject to pre-release modification, leading to inconsistencies between tables. Counts from census tables used in the synthetic reconstruction process have been modified as necessary to agree with one another. This has been achieved by selecting one of the published census distributions and using iterative proportional fitting to adjust all other small-area constraints in which the distribution features to agree with it.

3.4 Innovations in Combinatorial Optimisation

In similar fashion to synthetic reconstruction, a number of refinements to the basic approach to combinatorial optimisation outlined in Sect. 3.2 were implemented as part of the research reported in this chapter. The key innovations are summarised below.

3.4.1 Validated Random Number Generation

Combinatorial optimisation requires the generation of a number stream randomly different to at least the sixth digit, in order to ensure an equal chance exists of picking each of the approximately 250,000 households in the SAR. This is a non-trivial requirement that not all commercially available pseudorandom number generators can meet. For this project, therefore, one innovation has been the formal checking of potential random number generators for fitness for purpose (Voas and Williamson 1998).

3.4.2 Sequential Table Fitting

A problem previously identified is that, when selecting household combinations from survey microdata, some constraining tabulations are easier to satisfy than others (Williamson et al. 1998). Two steps have been taken to address this problem. First, tables based on 10% samples of census respondents have been replaced with modelled estimates of the full 100% sample distribution (as reported above). Second, an amended household selection routine has been tested, in which the hardest to fit tabulations were identified, based on the achieved tabular fit after a small number (e.g. 5,000) of household swaps. These hardest to fit tables were then used as constraints on household selection. Once satisfied, additional tables are reintroduced as constraints, with the added restriction that no changes in household combinations were allowed that impacted adversely on the level of fit already obtained for the hardest to fit tables (Voas and Williamson 2000a).

3.4.3 Stratified Household Selection

As originally implemented, combinatorial optimisation allowed any combination of households from the SAR to be selected that best satisfied small-area constraints. Two alternative approaches have also been evaluated, in which households can be selected only if they are from the same SAR region or from a ward with the same geodemographic ward classification (Wallace et al. 1995) as the small area being fitted.

3.4.4 RSSZ: A New Selection Criterion*

When selecting the set of households that best fit small-area constraints, the statistical measure of fit originally adopted was Total Absolute Error (TAE). However, TAE is poor at capturing relative error. The estimates presented in this chapter were therefore constrained to fit local area constraints using an alternative statistical measure, RSSZ*, based on the use of a modified *Z*-score, *Z**. (Sect. 3.6.1 provides further details.)

3.4.5 Stopping Rules

As finally implemented, an initial two million household combinations are evaluated for each small area. If even one cell in a single constraining table is deemed to be poorly fitted, up to a further two million evaluations are undertaken. At this stage, a small number of areas, comprising highly atypical households, still remain poorly fitted. In these cases, a further round of household replacement occurs (0.5 million evaluations), with potential replacement households restricted to those already found within the household combination. This strategy reflects the observation that, by the end of conventional household selection, the household combination contains a high concentration of the relevant atypical household types.

3.5 Understanding Between-Area Variation

Whatever method is adopted for creating synthetic microdata, the constraints to be met are supplied by small-area census tabulations. Resource constraints, in both person-hours and computing power, mean that not all of the available small-area constraints can be incorporated into the synthetic microdata generation process. It is necessary, therefore, to identify the minimum set of census counts that best capture between-area heterogeneity.

3.5.1 Spatial Concentration

To better understand the nature of between-area variation, an analysis was undertaken of the spatial scale of socio-economic variation across England and Wales, based on an analysis of 54 census variables chosen to reflect the full range of census topic coverage from the 1991 UK Census (Voas and Williamson 2000b). This analysis used the smallest UK Census output area for which 1991 Census data were released, the enumeration district (ED), with an average population of approximately 250 persons. It showed that, at ED level, ethnicity, dwelling type, housing tenure, transport mode, central heating, lone parenthood, qualifications and socio-economic group were the most spatially variable population attributes – that is, those displaying the greatest levels of within-area concentration (homogeneity) and, hence, between-area diversity (heterogeneity). Conversely, children (<16), persons aged 16–24, marital status, female employment, skilled or inactive males and long-term illness were the least spatially variable. When reanalysed at ward and district level, the basic ranking of the 54 census variables analysed remained similar, if not identical. (In crude terms, 1,000 EDs = 30 wards = 1 district). More specifically, for this analysis, spatial variability was measured using a dissimilarity score, *D*, adjusted to allow for the problems posed when analysing the distribution of rare populations across small areas (Voas and Williamson 2000b). The correlation coefficients of D when comparing wards with districts were 0.96, 0.96 for EDs vs. wards and 0.86 for EDs vs. districts.

However, this overall correlation masks the degree to which each geographic level influences overall spatial variability for a given variable. For example, over 80% of the spatial variability observable in ethnic concentrations at ED level is already observable at district level (i.e. ethnicity displays higher levels of heterogeneity between districts that between EDs within districts). Other variables with high levels of district-level heterogeneity are transport to work, access to cars, industry of employment and self-employment. The largest ward-level effects were found in professional occupations, qualifications, socio-economic groups and student concentration. Finally, some variables were found to be principally segregated only at ED level, including age and household size. These variations reflect the differing scales at which various social processes operate, including the labour and housing markets.

3.5.2 Multicollinearity

It might be tempting to focus efforts on accurately modelling the most spatially concentrated variables, as these appear to drive differentiation between small areas. However, such an approach would fail to take account of multicollinearity. If two variables are highly correlated, it may be necessary to accurately model only one of them, as the value of the second would be given by the value of the first.

The question is how small a set of variables would be necessary to adequately capture information about a core set of desired target variables $\{Y\}$. Analysis reveals that few variables can be left out if a high proportion of the overall variance is to be controlled for (Voas and Williamson 2001a). Even if an approximately optimal set of 25 variables is used to predict the value of each of the 29 remaining variables in the aforementioned census dataset, the average coefficient of determination (r^2) is only 63%. A simple explanation for these findings is that one-sixth of the 54 variables considered are not strongly associated with each other at ED level ($-0.5 < r < +0.5$).

An alternative solution is to attempt to reduce the 54 selected census variables into a more limited set of dimensions using principal components analysis. But even this statistically more sophisticated approach fairs little better. The first four components jointly account for just over half (54%) of ED-level variation. Twenty-five components are required to capture 90% of the observed variation. One implication is that approaches to area classification based upon data reduction, such as geodemographic profiling, are likely to only poorly summarise between-area differences. For synthetic estimation, the implication is that all variables of interest need to be directly modelled.

3.6 A Framework for Validating Small-Area Microdata

Previous evaluations of the quality of synthetic population microdata have been fairly rudimentary (Birkin and Clarke 1988; Williamson et al. 1998; Duley 1989). Consideration has been given to the fit of microdata to at most one or two published small-area tabulations, despite the use of multiple constraining tabulations. Such evaluations were undertaken only at small-area level, with no consideration of possible biases that might emerge during aggregation to larger geographical units. In addition, measures of fit were confined to the application of Z-tests to constraining cell counts. For the evaluation reported in this chapter, a far more extensive framework for validating small-area microdata was developed (Voas and Williamson 2001b; Huang and Williamson 2001b). A summary of the framework developed is given below.

3.6.1 *Identification of Appropriate Measures of Fit*

A review of a dozen statistical measures concluded that the most suitable for assessing the fit of synthetic microdata to published small-area constraints was the normal *Z*-score and related variants (Voas and Williamson 2001b). The normal *Z*-score has the advantage of familiarity, relative ease of calculation and the ability to identify both distributional and absolute errors (unlike Total Absolute Error, which focuses solely on absolute error).

In the context of synthetic microdata estimation, where E_{ij} = the expected small-area count for cell i in constraint table j, S_{ij} = the synthetic microdata estimate for this cell, N_j = the total count in table j, where p_{ij} = the expected proportion of counts falling in this cell, E_{ij}/N_j, and t_{ij} = the synthetically estimated proportion of counts falling in this cell, S_{ij}/N_j

$$Z_{ij} = \frac{\left(t_{ij} - p_{ij}\right)}{\sqrt{\dfrac{p_{ij}\left(1 - p_{ij}\right)}{N_j}}}.$$

If the *Z*-score for the difference between a synthetic and target cell count exceeds the relevant 5% *Z*-score critical value, then that cell is judged to be a 'non-fitting cell' (NFC). As a 'non-fitting' cell might be attributable at least in part to statistical disclosure control, a second less stringent definition of cellular fit was also used. A 'poorly fitting' cell (PFC) is a synthetic count that fails to fit even the published count ±1. In addition, ΣZ^2, the sum of the squared Z-scores for the n cells in a table, has a χ^2 distribution with n degrees of freedom, allowing tabular fit to be assessed. If ΣZ^2 for a given table is greater than the relevant 5% χ^2 critical value, then it is judged to be a 'non-fitting table' (NFT). Finally, dividing the squared Z-scores for a table by the appropriate 5% χ^2 critical value gives an additional measure, RSSZ, the relative sum of squared Z-scores, which when summed across all tables provides a measure of overall fit.

Ideally, RSSZ would replace TAE as the measure of fit used to help drive household selection during the process of combinatorial optimisation. Unfortunately, the calculation of Z_{ij} assumes that the expected and synthetic table totals match. When the synthetic and expected table totals differ markedly, Z-scores can remain low because the relative distribution of synthetic and expected counts is still similar. For this reason, a modified version of Z, Z*, is preferred for use in driving household selection. For Z*, the table total table used to calculate p_{ij} and s_{ij} is based upon the *expected* (i.e. small-area constraint) table total N_j^*, giving $t^*_{ij}=E_{ij}/N_j^*$, $p^*_{ij}=S_{ij}/N_j^*$ and

$$Z^*_{ij} = \frac{\left(t^*_{ij} - p^*_{ij}\right)}{\sqrt{\dfrac{p^*_{ij}\left(1 - p^*_{ij}\right)}{N^*}}}$$

Note that when synthetic and target table totals converge, Z* = Z.

Z* offers the additional computational advantage; when used, via RSSZ*, as a driver of household swaps, the numerator of Z_{ij}^* needs recalculating when a household swap affects the relevant synthetic cell count.

Self-evidently, the values of both Z and Z* remain undefined when the target small-area count is 0 and the synthetic count is greater than 0 ($p_{ij}=0/N_j$). In such circumstances, a mathematical justification has been provided for the practise of defining Z (or Z*) as equal to the equivalent synthetic count, S_{ij} (Voas and Williamson 2001b).

3.6.2 Innovations in Types of Fit Measured

Sections 3.7 and 3.8 assess the performance of synthetic microdata estimated using synthetic reconstruction and combinatorial optimisation. These assessments involve five main areas of innovation in the types of goodness-of-fit used. First, as already noted, the use of ΣZ^2 allows assessment for the first time of both tabular and overall fit to known constraints (NFT and RSSZ). Second, an attempt is made to allow for the impacts of pre-release census data modification (statistical disclosure control) when assessing levels of cellular fit (PFC). Third, multiple synthetic populations are created in acknowledgement of the inherently stochastic nature of both synthetic reconstruction and combinatorial optimisation. This allows both the average error (mean fit) and the bias (fit of the mean) of each estimation approach to be assessed. The range within which 95% of synthetic values lie is also calculated, giving a measure of variance. Fourth, consideration of fit is extended to include not only variable interactions fully constrained during the estimation process but also partially constrained and wholly unconstrained interactions. Fifth, the impact of spatial aggregation on the fit of synthetic microdata is evaluated.

3.7 The Impact on Combinatorial Optimisation of Selected Improvements

Section 3.4 highlighted a number of key technical improvements to the combinatorial optimisation algorithm. In this section, the relative performance advantage offered by these improvements is assessed. The test bed for this evaluation is the synthetic microdata generated, via combinatorial optimisation, for the 86 enumeration districts comprising the suburban Cookridge and inner-city University wards of Leeds. The sample survey used was the 1991 Household Sample of Anonymised Record – a 1% public use microdata set released following the 1991 Census. Small-area constraints to the estimation process were supplied by a set of 9 small-area tabulations published as part of the output from the 1991 UK Census. These small-area constraints are listed in Table 3.1. All 10% census counts were replaced with modelled 100% counts.

*3.7.1 Substitution of TAE with RSSZ**

Table 3.2 illustrates the impact on combinatorial optimisation of changing the household selection criteria from TAE to RSSZ*. Results are presented for three example enumeration districts. DAGF04 and DAGF12 are inner-city EDs with highly atypical population compositions, falling outside the 98th and 99.8th percentile of EDs, respectively, when ranked by difference from the national average across 54 selected census variables. ED DAFJ01 is a typical suburban ED lying relatively close to the national average.

Using improvements in TAE to guide household selection, at least one constraining table does not fit for ED DAGF12 (NFT>0), no matter how many evaluations are performed. With the use of RSSZ*, all the constraining tables are satisfied within 100,000 evaluations. At the cellular level, the performance advantage of RSSZ* is even greater, with only one or two cells out of 597 having *Z*-scores exceeding their critical values (NFC>0), compared to more than 17 if TAE is used to guide selection. Similar but less dramatic gains are observed for the suburban ED DAFJ01. The gains in algorithmic efficiency more than compensate for the 40% increase in calculation time (CPU seconds) per evaluation for RSSZ* as compared to TAE. These gains, attributable to an improved focus on relative rather than absolute fit, also

Table 3.2 Results from the use of TAE and RSSZ* as the selecting criterion

Selection criterion	TAE					RSSZ*				
Evaluations ('000)	TAE	RSSZ	NFT	NFC	CPU (s)	TAE	RSSZ	NFT	NFC	CPU (s)
(A) ED DAFJ01 in Cookridge ward (198 households)										
0	1,438	124.60	9.0	117.8	0	1,438	124.60	9.0	117.8	0
10	447	7.51	1.2	28	0	495	2.71	0	15.8	0
100	188	1.26	0	6.4	2	185	0.52	0	0.2	3
500	145	0.86	0	3.4	9	111	0.30	0	0	13
1,000	135	0.82	0	3.6	19	97	0.27	0	0	26
1,500	118	0.73	0	2.6	28	93	0.26	0	0	40
2,000	107	0.64	0	2.2	38	86	0.24	0	0	53
2,500	102	0.67	0	2.6	47	81	0.24	0	0	66
3,000	101	0.65	0	2.4	57	81	0.23	0	0	79
3,500	98	0.63	0	1.8	66	78	0.23	0	0	92
4,000	97	0.60	0	1.4	75	80	0.23	0	0	105
5,000	94	0.60	0	1.4	94	77	0.22	0	0	131
6,000	93	0.60	0	1.4	113	76	0.22	0	0	158
8,000	91	0.59	0	1.4	151	74	0.21	0	0	210
10,000	89	0.57	0	1.2	188	73	0.21	0	0	263
(B) ED DAGF04 in University ward (149 households)										
0	1,869	48.89	9.0	132.6	0	1,869	48.89	9.0	132.6	0
10	880	12.11	5.6	67.0	0	853	8.14	3.0	59.2	0
100	364	4.11	0	21.8	2	359	1.88	0	4.6	3
500	320	3.57	0	18.8	9	236	0.98	0	0.6	13
1,000	275	3.07	0	14.8	19	206	0.81	0	0.4	26
1,500	248	2.72	0	13.2	28	200	0.75	0	0.4	39
2,000	240	2.58	0	13.8	38	190	0.70	0	0.2	53
2,500	233	2.35	0	13.0	47	185	0.67	0	0.2	66
3,000	227	2.22	0	12.0	57	178	0.65	0	0.2	79
3,500	222	2.16	0	11.8	66	173	0.62	0	0.2	92
4,000	219	2.18	0	10.4	75	177	0.62	0	0.2	105
5,000	216	2.16	0	10.4	94	171	0.60	0	0.2	131
6,000	213	2.11	0	10.6	113	162	0.58	0	0.2	158
8,000	207	2.03	0	10.0	151	160	0.56	0	0	210
10,000	201	1.98	0	9.8	188	153	0.53	0	0	263
(C) ED DAGF12 in University ward (191 households)										
0	2,642	106.81	9	164.6	0	2,642	106.81	9	164.6	0
10	1,542	35.15	8.6	116.8	0	1,421	17.24	7.8	103.2	0
100	659	7.56	3.4	43.6	2	680	3.97	0	19.0	3
500	445	5.39	1.8	23.2	9	398	1.59	0	4.2	13
1,000	385	4.29	1.0	20.6	19	343	1.24	0	3.0	26
1,500	355	3.98	1.2	18.8	28	315	1.11	0	2.0	40
2,000	338	3.83	1.2	16.6	38	295	1.05	0	2.2	53
2,500	324	3.87	1.4	17.4	47	294	1.02	0	1.6	66
3,000	314	3.69	1.2	17.8	57	286	0.98	0	1.6	79

(continued)

Table 3.2 (continued)

Selection criterion	TAE					RSSZ*				
Evaluations ('000)	TAE	RSSZ	NFT	NFC	CPU (s)	TAE	RSSZ	NFT	NFC	CPU (s)
3,500	309	3.69	1.2	17.4	66	284	0.95	0	1.4	92
4,000	305	3.69	1.2	18.0	76	278	0.95	0	1.4	105
5,000	300	3.61	1.2	16.6	94	271	0.90	0	1.6	132
6,000	296	3.64	1.4	17.8	113	269	0.88	0	1.4	158
8,000	293	3.57	1.2	17.2	151	269	0.86	0	1.2	211
10,000	290	3.54	1.2	17.6	189	261	0.85	0	1.2	263

Figures are 5-run average
Total number of tables 9; Total number of cells 597
CPU time is central processing unit time in seconds on a 800 MHz PC

allow RSSZ*-driven combinatorial optimisation to replace the sequential table fitting necessary when using TAE (c.f. Sect. 3.4.2) with the simultaneous table fitting approach outlined in Sect. 3.2.

3.7.2 *Stratified Household Selection*

As originally conceived, combinatorial optimisation selected the households chosen to represent a small area from the whole of the SAR dataset (W). An alternative strategy (R) is to select households drawn from only the same SAR region as the small area being estimated. As Fig. 3.2 illustrates, for suburban ED DAFJ01 restriction of sampling to a regional subset of the SAR leads to only a slight deterioration in performance, but for inner-city ED DAGF12, the outcome is significantly worse. The results suggest that for atypical EDs-limiting household selection to region-specific SAR significantly increases the error of estimation. The strategy finally adopted ($R+W$) initially restricts household selection to region-specific SAR, but this restriction is lifted after evaluation of 200,000 household combinations if one or more constraining cells remain to be fitted. Using this strategy, at the end of ten million evaluations, 100% of households selected to represent ED DAFJ01 are drawn from the relevant regional SAR, compared to 14% of households for ED DAGF12.

An alternative considered but subsequently discarded was to restricting selection to households drawn from wards of the same geodemographic type as the small area being estimated (Voas and Williamson 2000a). The potential gains for this approach are necessarily limited by the weaknesses inherent in all area classifications (reviewed in Sect. 3.2). It was found that selecting households on the basis of area type led to a marked improvement in fit for some but not all population attributes, leading to the conclusion that a better strategy would be simply to increase the number of constraints on the estimation process.

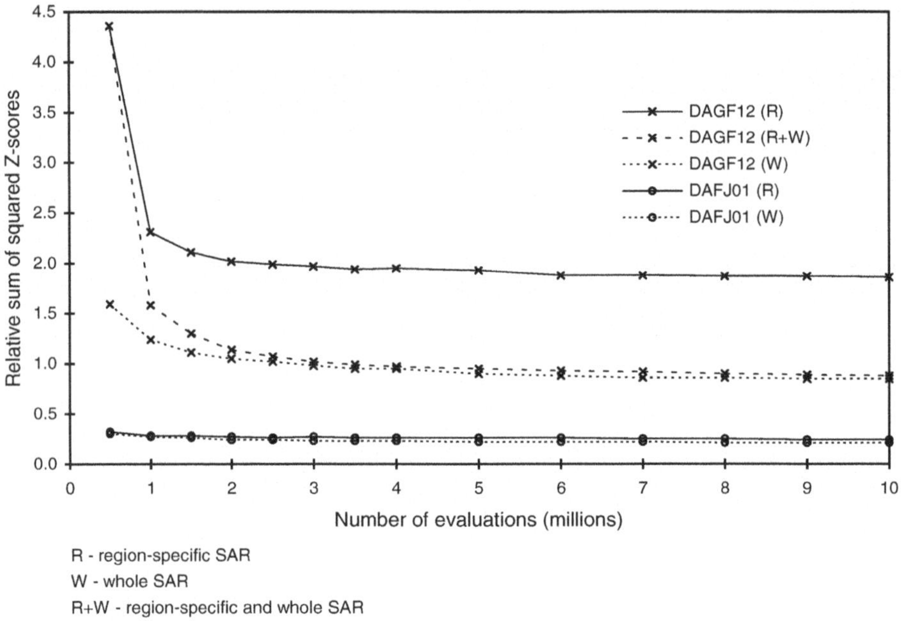

Fig. 3.2 Comparing the use of region-specific vs. whole SAR

3.8 Synthetic Reconstruction vs. Combinatorial Optimisation

For comparison with the combinatorial optimisation outputs already introduced (POP91CO), synthetic microdata were also generated by synthetic reconstruction (POP91SR) for the same 86 enumeration districts, using the same survey data and local area constraints as inputs. Additional information from higher-level geographies was incorporated in the synthetic reconstruction process in order to support robust estimation of the required conditional probabilities. However, lack of small-area data restricted synthetic reconstruction to the allocation of ethnic group for household heads only. Consequently, when assessing the relative performance of the two approaches, the fit to those tables involving a whole-population ethnic breakdown is disregarded. Finally, to allow for an assessment of the impact of random variability, 100 runs of each approach were undertaken.

3.8.1 ED-Level Mean Fit

The fit of each set of synthetic microdata to published small-area counts and tabulations has been evaluated using measures based, respectively, on *Z*-scores and ΣZ^2 (as outlined in Sect. 3.6).

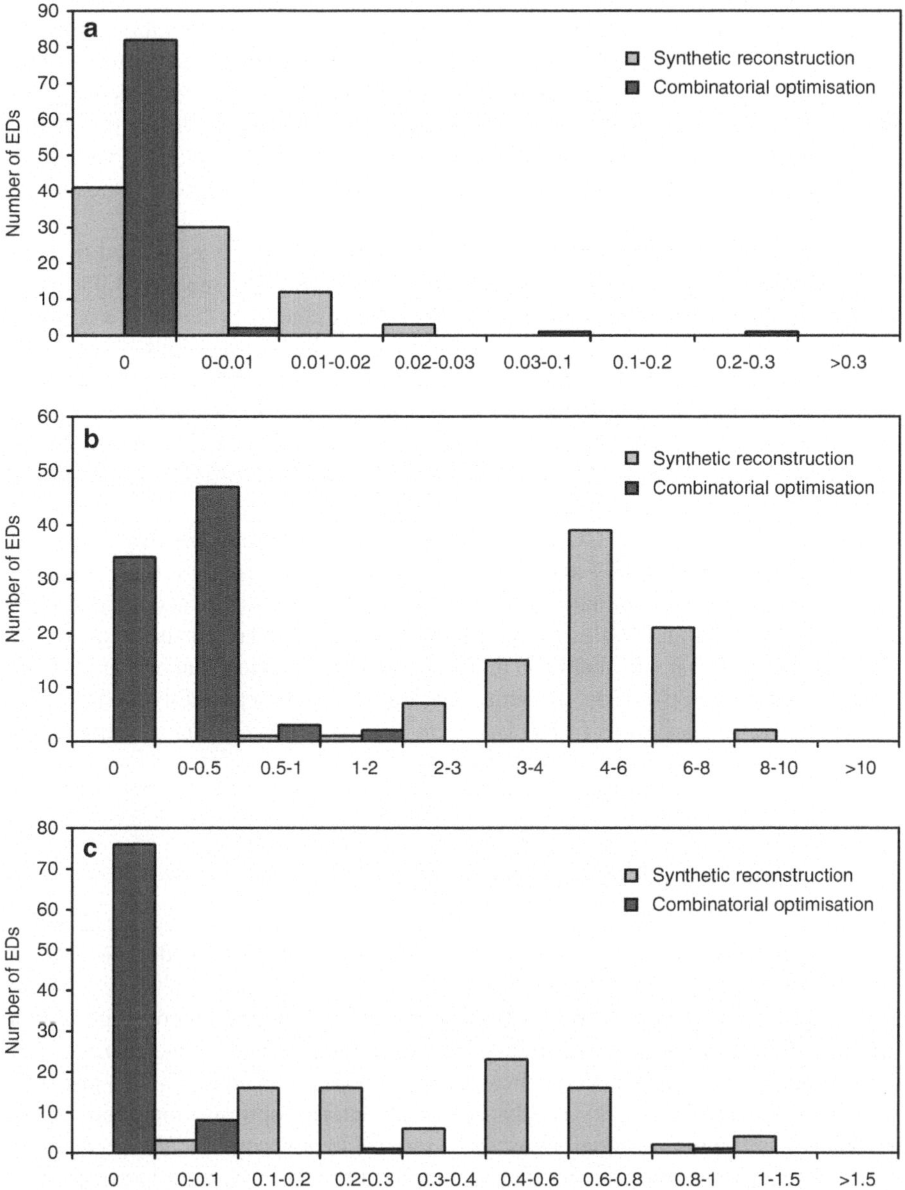

Fig. 3.3 Performance comparison: synthetic reconstruction vs. combinatorial optimisation. (**a**) Histogram of mean number of NFT over 100 replications. (**b**) Histogram of mean number of NFC over 100 replications. (**c**) Histogram of mean number of PFC over 100 replications

To assess the general accuracy of the two estimation strategies, the fit of 100 replications per ED was calculated separately and the mean taken, giving the mean fit. Figure 3.3 presents the distribution of this mean fit per ED, at both tabular and cellular level. As shown in Fig. 3.3a, for nearly half of all EDs in the test area, synthetic reconstruction (POP91SR) led to synthetic datasets with a mean number of NFT equal to zero (i.e. the synthetic data fit all constraining tables for all 100 trials). The figures for the remaining EDs are all less than 0.03, equivalent to one table (out of 7) in three trials (out of 100) failing to fit. The tabular fit for combinatorial optimisation (Pop91CO) is even better. For all but four EDs, the NFT values are zero. Only in two EDs is the fit produced by Pop91CO less good than that of Pop91SR.

At cellular level, the number of non-fitting cells for datasets generated by synthetic reconstruction (Pop91SR) ranges from 0.78 to 9.18 with a mean of 4.95 across all 86 EDs (Fig. 3.3b). This result means that, on average, only 5 out of a possible 415 cells fail the *Z*-test in a given trial. Allowance for the ±1 uncertainty over actual cell values leads to a tenfold reduction in poorly fitting cells (Fig. 3.3c). On average, the number of PFC is only 0.45, that is, less than one cell poorly fitted per trial. Combinatorial optimisation (Pop91CO), however, produces an even better fit at the cellular level (Fig. 3.3b, c). The average number of NFC and PFC over 86 EDs is only 0.13 and 0.02, respectively.

The two EDs that Pop91CO fails to produce better estimates for than Pop91SR are the student EDs DAGF57 and DAGF58. These two EDs are extremely atypical; their distance from the national average has been identified as the second and third highest in England and Wales (Voas and Williamson 2000a). In these cases, the required households are probably so unusual that no equivalents can be found in the SAR. Even for these two EDs, the actual test statistics may reasonably be described as very good; the hardest to fit ED DAGF58 averages only 0.29 non-fitting tables and 1.73 non-fitting cells per replication.

3.8.2 ED-Level Fit of the Mean

As well as assessing the mean fit over 100 runs, it is possible to consider the 100-run mean of the synthetically estimated cell values and assess the fit of these 100-run mean counts to the target census-based small-area constraints, thereby giving an indication of overall bias. For synthetic reconstruction (Pop91SR), the average fit of the 100-run means at ED level (as measured by RSSZ of the mean) is 0.13 and 0.18 for Cookridge and University wards. For combinatorial optimisation (Pop91CO), the equivalent figures are equally low, at 0.11 and 0.23, respectively. The greater gap in RSSZ between wards indicates that the output of Pop91CO is slightly more sensitive to location that Pop91SR.

The strengths and weaknesses of the two approaches are summarised in Table 3.3, which presents estimated counts for one of the constraining tables, a cross tabulation of sex by marital status. Although the mean synthetic counts are more or less identical across the two simulation methodologies, the range of synthetic counts over 100 replications is far greater for synthetic reconstruction than for combinatorial optimisation. Table 3.3 also helps to make clear the meaning of 'fit'. As can be

Table 3.3 Comparing the fit of estimated population for ED DAFJ01 to SAS Table 34

		Synthetic reconstruction				Combinatorial optimization			
(a) *Cellular test*	SAS Table 34	Mean synthetic	Top and bottom of 95% interval		% of |Z|>1.96	Mean synthetic	Top and bottom of 95% interval		% of |Z|>1.96
Male, single, widowed, divorced									
Employees – full time	17	18.8	23	14	0	18.0	19	17	0
Employees – part time	4	4.2	8	2	3	4.0	4	4	0
Self-emp. – with employees	0	0.0	0	0	0	0.0	0	0	0
Self-emp. – without employees	4	4.5	7	3	2	4.0	5	4	0
On a government scheme	1	0.0	0	0	0	1.0	1	1	0
Unemployed	1	1.0	3	0	5	1.1	2	1	0
Students	8	8.3	12	5	0	8.0	8	7	0
Permanently sick	1	1.1	3	0	8	1.0	1	1	0
Retired	12	12.9	15	11	0	12.0	13	11	0
Other inactive	1	0.9	2	0	1	1.0	1	1	0
Male, married									
Employees – full time	63	63.4	68	60	0	62.3	64	61	0
Employees – part time	10	10.1	14	8	0	9.8	10	9	0
Self-emp. – with employees	11	10.9	15	7	0	9.7	10	9	0
Self-emp. – without employees	6	6.1	9	3	0	6.2	7	6	0
On a government scheme	0	0.0	0	0	0	0.0	0	0	0
Unemployed	1	1.0	3	0	7	1.7	2	1	0
Students	1	0.7	2	0	1	1.0	1	1	0
Permanently sick	6	5.5	8	3	0	6.0	7	5	0
Retired	41	41.0	44	38	0	41.1	42	40	0
Other inactive	0	0.0	0	0	0	0.0	0	0	0
Female, single, widowed, divorced									
Employees – full time	21	21.5	26	18	0	20.6	21	20	0
Employees – part time	10	10.6	15	6	0	9.8	11	9	0

Table 3.3 (continued)

(a) *Cellular test*	SAS Table 34	Synthetic reconstruction: Mean synthetic	Synthetic reconstruction: Top and bottom of 95% interval		Synthetic reconstruction: % of |Z|>1.96	Combinatorial optimization: Mean synthetic	Combinatorial optimization: Top and bottom of 95% interval		Combinatorial optimization: % of |Z|>1.96
Self-emp. – with employees	0	0.0	0	0	0	0.0	0	0	0
Self-emp. – without employees	0	0.0	0	0	0	0.0	0	0	0
On a government scheme	0	0.0	0	0	0	0.0	0	0	0
Unemployed	0	0.0	0	0	0	0.0	0	0	0
Students	10	10.5	13	8	0	10.7	11	10	0
Permanently sick	1	1.1	3	0	4	1.0	1	1	0
Retired	17	18.1	21	16	0	17.7	18	17	0
Other inactive	9	9.5	13	6	2	9.3	10	9	0
Female, married									
Employees – full time	23	22.3	26	19	0	22.9	24	22	0
Employees – part time	39	38.3	43	33	0	38.5	39	37	0
Self-emp. – with employees	2	2.0	4	0	0	1.4	2	1	0
Self-emp. – without employees	7	7.0	10	4	0	6.2	7	6	0
On a government scheme	0	0.0	0	0	0	0.0	0	0	0
Unemployed	2	1.7	4	1	0	1.0	1	1	0
Students	0	0.0	0	0	0	0.0	0	0	0
Permanently sick	3	3.0	6	1	0	3.0	4	3	0
Retired	32	31.2	34	28	0	31.5	33	30	0
Other inactive	34	32.9	38	29	0	33.7	35	33	0

(b) *Tabular test*	Synthetic reconstruction	Combinatorial optimisation
SSZ of mean	1.8	1.6
Mean TAE	37.2	11.9
Mean SSZ	12.8	2.4
% of SSZ>critical value[a]	0	0
Mean NFC	0.3	0
Mean PFC	0	0

[a]5% chi-square critical value=55.8; Number of replications=100

seen, for combinatorial optimisation, the mean synthetic counts are extremely close to the constraining census counts, as are the synthetic counts at the top and bottom of the trimmed 95% estimate range.

3.8.3 Ward-Level Fit

Superior performance at the ED level does not necessarily guarantee superior performance when synthetic populations are aggregated to ward level. As Table 3.4 demonstrates, at ward level, the overall fit of the mean of the 100 Pop91SR estimates (RSSZ of mean) is closer to the target distribution than that for Pop91CO. However, in almost every other respect, Pop91CO continues to outperform Pop91SR, in particular offering markedly reduced overall levels of variance (lower average numbers of non-fitting cells and tables). A similar story is found when the fit to individual constraining tables is considered. For most purposes, only a single set of synthetic microdata will be used. Therefore, a guaranteed close fit (minimal variability) is to be preferred to assurances of minimum bias over 100 trials.

3.8.4 Fit of Unconstrained Counts

So far, the focus of this evaluation has been on how well the synthetic microdata fit the target small-area counts used to constrain their estimation. As Table 3.2 shows, these estimation constraints include a number of local interactions between variables, such as the interaction between age, sex and economic position. However, not all of the possible interactions between variables in the synthetic microdata have been constrained during the estimation process. For example, whilst the local area distributions of socio-economic and marital status have been constrained, the interaction between these two variables has not. Given that the margins have been constrained, this type of interaction might best be described as 'partially constrained'. An alternative scenario is one in which none of the variables involved in the interaction have been constrained in any way as part of the estimation process. This type of interaction might be described as 'fully unconstrained'.

A second test for synthetic microdata, therefore, is how well they capture these unconstrained interactions between variables used in constraining tables. This problem has been considered for combinatorial optimisation outputs. Standard census outputs contain insufficient overlap in variables to allow for assessment of the fit of synthetic microdata to a wide range of partial and wholly unconstrained interactions. Therefore, artificial enumeration districts, of average population size, were created via stratified sampling of households from the SAR (Voas and Williamson 2000a). Two of the artificial EDs created, 'Middle England' and 'Rural', have a population composition very close to the national average. The 'Deprived industrial' ED is as far from the norm as the suburban ED DAFJ01, whilst the 'Deprived urban, council flats' ED is as highly atypical as the inner-city ED DAGF04, where

Table 3.4 Performance of synthetic reconstruction and combinatorial optimisation at ward level

(a) Overall fit	Cookridge ward Overall TAE	Overall RSSZ	Number of NFT	Number of NFC	RSSZ of mean	University ward Overall TAE	Overall RSSZ	Numberm of NFT	Number of NFC	RSSZ of mean
Synthetic reconstruction	2,307	2.98	0.17	15.6	0.44	2,701	3.64	0.8	19.3	0.31
Combinatorial optimisation	1,084	0.84	0	2.3	0.64	1,498	1.28	0	1.9	1.05

(b) Tabular fit			Cookridge ward					University ward				
Table	Number of cells	Critical value	TAE	SSZ and % of SSZ > critical value		Number of NFC	SSZ of mean	TAE	SSZ and % of SSZ > critical value		Number of NFC	SSZ of mean
Synthetic reconstruction												
39	28	41.3	6	0	0	0	0	6	0.1	0	0.0	0.1
35	68	88.3	687	67.5	14	4.3	3.3	836	108.4	74	7.7	3.8
34	40	55.8	401	29.6	3	1.1	2.8	399	28.1	2	0.9	1.8
8	180	212.3	768	141.1	0	8.4	40.2	835	161.7	4	8.9	40.6
42	7	14.1	85	3.0	0	0.1	0.2	112	4.3	0	0.1	0.1
49	16	26.3	99	12.7	0	0.5	3.7	164	11.7	0	0.3	0.8
86	76	97.4	262	31.8	0	1.2	1.4	348	38.1	0	1.4	0.9
Combinatorial optimisation												
39	28	41.3	50	0.6	0	0	0.4	68	1.0	0	0	0.6
35	68	88.3	240	7.8	0	0	5.5	273	12.3	0	0.0	8.9
34	40	55.8	197	4.8	0	0	5.3	260	12.9	0	0.2	14
8	180	212.3	393	60.3	0	1.7	39.1	505	86.4	0	1.6	60.2
42	7	14.1	40	0.9	0	0	0.7	100	3.2	0	0	2.9
49	16	26.3	28	2.8	0	0	2.1	75	2.3	0	0	1.8
86	76	97.4	138	18.9	0	0.7	14.9	218	15.7	0	0.1	12.1

Critical values are table-specific 5% critical values (degrees of freedom = number of cells)
Test statistics are averages over 100 replications. Number of cells in all tables = 415

Table 3.5 Fit of synthetic microdata to partially and fully constrained cross tabulations

	Artificial EDs			
	(Samples of SAR households stratified by ONS ward type, region and tenure)			
	Strata			
	Rural South West Any tenure	'Middling England' East Midlands Any tenure	Deprived industrial North Any tenure	Deprived urban Outer London Council flats
Cross tabulations[a]	% of 'non-fitting' tables (over 200 replications)			
'Partially unconstrained' tables				
Socio-economic group/household composition	0	0	0	0
Socio-economic group/no. of rooms	0.5	0	0	0
Household composition/dependants	0	0	0	0
Dependants/tenure	0	0	0	0
Sex/marital status/tenure	16.0	1.5	1.5	0
Long-term illness/sex	0	1.5	0	0
'Wholly unconstrained' tables				
Qualifications/age/sex	14	48	20	1
Migration/age	27	77	60	76
Car availability/adults	87	52	96	31
Ethnic group/country of birth	99	67	100	100
Life stage/couple household	24	55	22	88
Household space type	100	64	100	100

[a]Italicised variables not present in tables used as constraints during synthetic microdata estimation

distance from the norm is measured by the Euclidean distance of each ED from the centroid of the 54 standardised census variables already discussed in Sect. 3.5. As Table 3.2 shows, the fit of five partially constrained tabulations, estimated using a set of 8 constraint tables (Voas and Williamson 2000a), was excellent and good (85% of runs fit) for the remaining tabulation of sex/marital status/tenure, which cuts across individual and household levels. In contrast, the fit on interactions between variables not involved in the constraining process (Table 3.5) is generally extremely poor. Similar results were found when assessing the fit of partially and totally unconstrained tabulations at ward level, for both synthetic reconstruction and combinatorial optimisation (Huang and Williamson 2001b).

3.9 Conclusion

Combinatorial optimisation is a superior approach to synthetic reconstruction for the generation of small-area microdata. In particular, combinatorial optimisation offers a marked reduction in variability of performance between runs. Wholly

Fig. 3.4 The estimated distribution of 'yuppies' in York

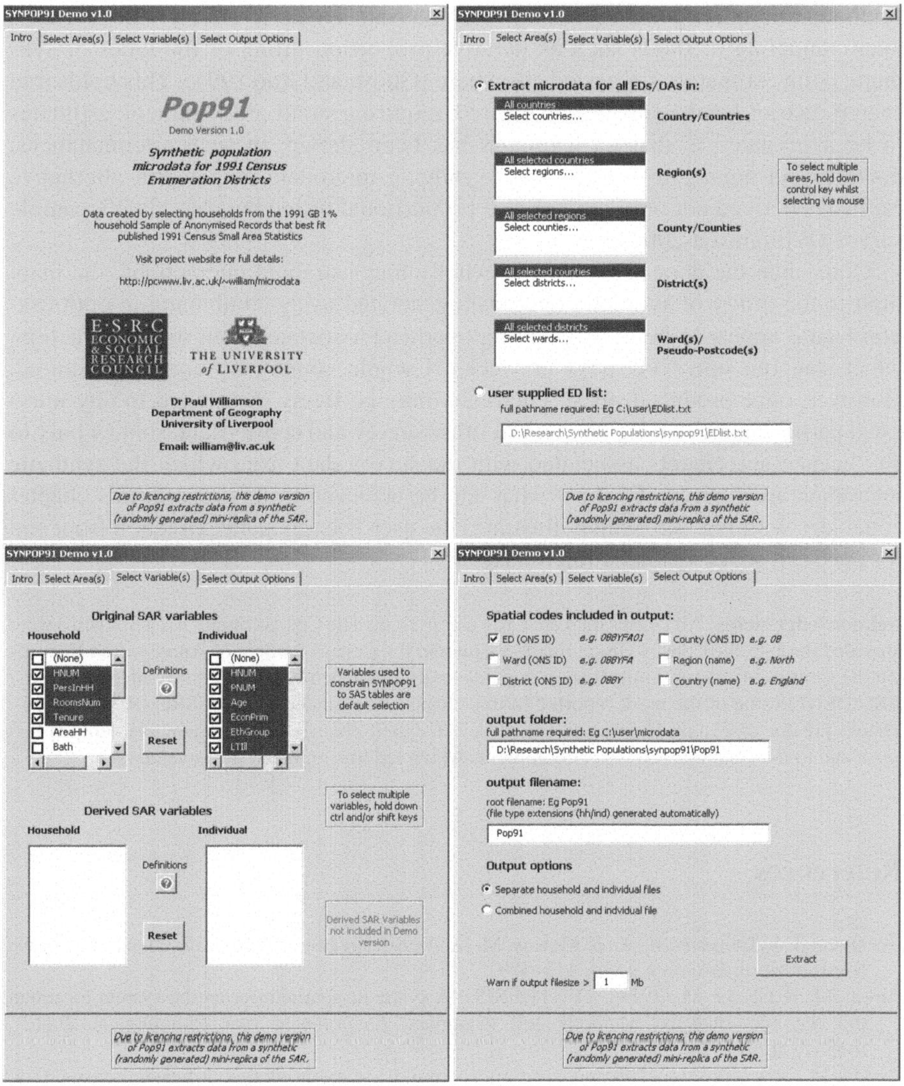

Fig. 3.5 Screenshots of user-friendly data extraction software

unconstrained interactions remain poorly captured, but partially constrained interactions in most cases provide good fit at the tabular level. As a result, synthetic microdata generated using combinatorial optimisation offer clear 'added value', providing reliable estimates of many previously unknown cross tabulations. An example is Fig. 3.4, which maps the estimated distribution of young urban professional ('yuppie') households in York (households containing only residents aged 18–34, headed by a 'professional' or 'manager'). For this reason, when flexibility of aggregation or estimates of large numbers of unknown tabulations are required, synthetic microdata generated using combinatorial appear to offer a better solution than competing methods such as iterative proportion fitting or the types of synthetic point estimators suggested elsewhere (Ghosh and Rao 1994). This holds true even if district-level estimates are desired, requiring small-area synthetic estimates to be aggregated to district level. As has been shown, in such circumstances, district-level aggregates of ED-level synthetic microdata outperform alternative estimates derived not only via iterative proportional fitting but also via 2% sample survey (Williamson 2007).

Other than the poor estimation of wholly unconstrained interactions, the main limit to the utility of synthetic microdata generated using combinatorial optimisation would appear to be the computing overhead associated with their production, which can run into CPU days or weeks if whole country coverage is required. However, once produced, these microdata may be freely distributed to any interested parties, in much the same way as other survey and census data, subject only to any licence agreements associated with the survey data from which the synthetic microdata are derived. To show what can be achieved in this regard, this chapter concludes with Fig. 3.5, which illustrates the user-friendly data extraction front end for one such set of synthetic microdata.

Acknowledgements All census data are Crown copyright. The Census Small-Area Statistics were provided through the Census Dissemination Unit and the Census Sample of Anonymised Records via the Census Microdata Unit, both at the University of Manchester and both funded by ESRC/JISC/DENI. Some of the work reported in this chapter was funded by the ESRC (R000237744). Thanks are due to Zengyi Huang and David Voas for their contributions to many elements of this work and to the chapter's referee for comments which led to significant improvements.

References

Beckman, R. J., Baggerly, K. A., & McKay, M. D. (1996). Creating synthetic baseline populations. *Transportation Research A, 30*, 415–429.

Birkin, M., & Clarke, M. (1988). SYNTHESIS – A synthetic spatial information system for urban and regional analysis: Methods and examples. *Environment and Planning A, 20*(12), 1645–1671.

Birkin, M., & Clarke, G. (1995). Using microsimulation methods to synthesize census data. In S. Openshaw (Ed.), *Census users' handbook* (pp. 363–388). Cambridge: GeoInformation International.

Clarke, G. (1996). Microsimulation: An introduction. In G. P. Clarke (Ed.), *Microsimulation for urban and regional policy analysis* (p. 3). London: Pion.

Dale, A., & Teague, A. (2002). Microdata from the Census: Samples of Anonymised Records (Chapter 14). In P. Rees, D. Martin, & P. Williamson (Eds.), *The census data system* (pp. 203–212). Chichester: Wiley.

Duley, C. J. (1989). *A model for updating census-based household and population information for intercensal years*. Unpublished PhD thesis, School of Geography, University of Leeds.

Ghosh, M., & Rao, J. N. K. (1994). Small area estimation: An appraisal. *Statistical Science, 9*, 55–93.

Huang, Z., & Williamson, P. (2001a). *A modified sampling procedure for small area population simulation* (Working Paper 2001/2). Liverpool: Population Microdata Unit, Department of Geography, University of Liverpool. (Available from: http://pcwww.liv.ac.uk/microdata)

Huang, Z., & Williamson, P. (2001b). *A comparison of synthetic reconstruction and combinatorial optimisation approaches to the creation of small-area microdata* (Working Paper 2001/2). Liverpool: Population Microdata Unit, Department of Geography, University of Liverpool. (Available from: http://pcwww.liv.ac.uk/microdata)

Martin, D., Nolan, A., & Tranmer, M. (2001). The application of zone-design methodology in the 2001 UK Census. *Environment and Planning A, 33*(11), 1949–1962.

Voas, D., & Williamson, P. (1998). *Testing the acceptability of random number generators* (Working Paper 1998/2). Liverpool: Population Microdata Unit, Department of Geography, University of Liverpool. (Available from: http://pcwww.liv.ac.uk/microdata)

Voas, D., & Williamson, P. (2000a). An evaluation of the combinatorial optimisation approach to the creation of synthetic microdata. *International Journal of Population Geography, 6*, 349–366.

Voas, D., & Williamson, P. (2000b). The scale of dissimilarity: Concepts, measurement and an application to socio-economic variation across England and Wales. *Transactions of the Institute of British Geographers, 25*, 465–481.

Voas, D., & Williamson, P. (2001a). The diversity of diversity: A critique of geodemographic classification. *Area, 33*(1), 63–76.

Voas, D., & Williamson, P. (2001b). Evaluating goodness-of-fit measures for synthetic microdata. *Journal of Geographical and Environmental Modelling, 5*(2), 177–200.

Wallace, M., Charlton, J., & Denham, C. (1995). The new OPCS area classification. *Population Trends, 79*, 15–30.

Williamson, P. (2002). Synthetic microdata. In P. Rees, D. Martin, & P. Williamson (Eds.), *The census data system* (pp. 231–241). Chichester: Wiley.

Williamson, P. (2007). Confidentiality and anonymised survey records: The UK experience. In A. Gupta & A. Harding (Eds.), *Modelling our future: Population ageing, health and aged care* (pp. 387–413). Amsterdam: Elsevier.

Williamson, P., Birkin, M., & Rees, P. H. (1998). The estimation of population microdata by using data from small area statistics and samples of anonymised records. *Environment and Planning A, 30*, 785–816.

Chapter 4
Estimating Small-Area Income Deprivation: An Iterative Proportional Fitting Approach

Ben Anderson

4.1 Background

As the chapters in this volume make clear, there is increasing demand for the development of small-area estimates of a range of socio-economic indicators – not only for research and public policy use but also for commercial applications. In the case of the former, indicators of exclusion and deprivation (Birkin and Clarke 1989; Williamson and Voas 2000; Gong et al. 2011; Tanton, Mcnamara et al. 2009; Tanton, Vidyattama et al. 2009) as well as ill health (Ballas et al. 2005b; Mohana et al. 2005; Smith et al. 2007; Morrissey et al. 2008; Edwards and Clarke 2009) and resource use (Williamson 2001; Druckman and Jackson 2008; Harding et al. 2011) are good examples. In the case of the latter, it is more usually wealth, consumption (expenditure) and lifestyle indicators that are of interest (Anderson 2008; Anderson et al. 2009a; Birkin and Clarke 2011; Vidyattama et al. 2011).

The need for local stakeholders, and especially policymakers, to understand the small-area distribution of deprivation has led to a number of approaches to the calculation or estimation of deprivation indicators at levels that cannot be robustly supported by current national surveys. This need has been most clearly met in the United Kingdom by the development of the Indices of Deprivation (IMD) series that now cover the constituent countries of England, Wales, Scotland and Northern Ireland (Noble et al. 2006).

Income deprivation has always been a key part of the IMD, and the income sub-domain provides measures of poverty through aggregation of benefits claimant counts (Noble et al. 2004, 2006, 2008). However, with the increasing focus of poverty-related policy on the standard income deprivation measure of the percentage of

B. Anderson (✉)
Centre for Research in Economic Sociology and Innovation,
Department of Sociology, University of Essex, Colchester, UK
e-mail: b.anderson@lancaster.ac.uk

R. Tanton and K.L. Edwards (eds.), *Spatial Microsimulation: A Reference Guide for Users*,
Understanding Population Trends and Processes 6, DOI 10.1007/978-94-007-4623-7_4,

households below a given income threshold (Gordon and Townsend 2000; Eurostat 2007), an alternative approach may be required (Noble et al. 2008).

Unfortunately, reliably and robustly estimating household and individual income at small-area levels in the United Kingdom is, and has always been, a challenging exercise. Although there have been a number of studies exploring the changing geography of deprivation, there is a general paucity of data relating to the small-area geography of household income, wealth and taxation.

The best source of small-area socio-economic information in the UK is the census of population, but unfortunately, in order to preserve confidentiality and minimise non-response, the UK census does not provide any information on variables, such as household income, wealth and taxation (Marsh 1993). Aspects of income are available from a range of government surveys, but in many cases, the finest spatial scale to which these survey data are coded is the 12 UK government office regions[1] or, in the case of some surveys, the local authority district, the main level of UK local administration with mean household population sizes of around 57,000. Even then, however, many such surveys provide incomplete geographical coverage, sampling only a fraction (and sometimes none) of the residents of any given district.

A method aimed at estimating the population of local households with incomes below a given threshold would ideally produce a national population distribution, which when aggregated would be within the known error bounds of the official estimates at the regional and national level; support the estimation of threshold-based income deprivation indicators; support the small-area level analysis of different policy outcomes; and be estimated at the lowest practical level of spatial geography.

Sensing this demand, social and economic geographers, as well as a number of commercial data providers, have developed a range of approaches to the estimation of such variables at small-area levels using both econometric (Gosh and Rao 1994; Rao 2003; Bates 2006) and spatial microsimulation approaches (Williamson 2005; Tanton et al. 2009a, b). This work has included a number of attempts to project small-area income distributions forwards in time (Ballas 2004; Ballas et al. 2005a), as well as Australian experimentation on techniques for ageing the spatial microdata (Harding et al. 2011; Vidyattama and Tanton 2010).

In this chapter, we briefly review some of these approaches before discussing a spatial microsimulation approach based on the iterative proportional fitting algorithm which can deliver against the characteristics listed above at the Welsh Lower Layer Super Output Area[2] (LSOA) level. This then allows us to firstly explore the value of including such estimates in official indicators of multiple deprivation and secondly enable comparison with the income domain scores from the Welsh IMD (WIMD) 2005 (Noble et al. 2006).

[1] See http://www.statistics.gov.uk/geography/gor.asp, mean size *c* 2 million households.

[2] The second level of census aggregation (containing multiple census 'output areas') containing on average around 600 households.

4.2 Small-Area Income Estimation Methods

Perhaps the simplest method of estimating small-area income levels is the use of variables available in the census as proxies for income based on correlations between these variables and income on national sample surveys. Thus, for example, the mean income for different socio-economic classifications can be calculated from a national income survey such as the Department for Work and Pensions' Family Resources Survey (FRS). Given that census data provides the number of people and/or household response persons within a given socio-economic classification, a simple multiplication would provide 'indirect non-survey-designed estimates' of income at geographical levels for which we have socio-economic information. However, such simplistic approaches take no account of more complex relationships between socio-economic variables and income distributions nor of the manner in which such relationships may vary in space.

Williamson and Voas (2000) used data from the large-scale UK Census Rehearsal of April 1999 which included a banded income question to test a range of household and area level predictors of small-area income levels (Williamson and Voas 2000). In this case, their analysis suggested that there was a high degree of income heterogeneity even at small-area level and that a relatively small number of indicators were the optimal predictors of area level income measures. Williamson suggested that by far, the most effective simple proxy for income is the proportion of the economically active population in National Statistics Socio-economic Classification (NS-SEC) categories 1 and 2 (managerial and professional occupations) and that this finding applies regardless of whether mean income is measured per person, per adult or per number of persons in the household. In addition, Williamson's analysis suggested that this indicator performed better than contemporary deprivation indices and interestingly that only 1% of unexplained between-adult income variation could be explained by area level factors such as house prices.

The Office for National Statistics has subsequently developed a regression-based approach to the estimation of small-area income levels (Heady et al. 2003; Bates 2006). Their method involves combining survey data with other data sources that are available on an area basis and is underpinned by the area level relationship between the survey and auxiliary variables such as administrative or census data. In this context, they modelled ten variables at the small-area level: household income from the Family Resources Survey, household income from the General Household Survey, a measure of social capital, the number of children from ethnic minorities, the number of people available to help in a crisis, the number of single-parent families, overcrowding and three measures of poor health. This small area estimation project (SAEP) methodology has been applied by the ONS to produce ward-level estimates of mean household income for 2001/2002 and middle layer super output area[3] level-level estimates for 2004/2005. Unfortunately, the approach has not yet been

[3] Administrative areas that are nested within local authorities and which are exact aggregates of, on average, five LSOAs.

applied to the estimation of threshold-based indicators such as the percentage of households below a given threshold, and it also has not been used to create estimates at lower levels of geography such as for LSOAs.

An alternative approach, generally implemented in the commercial market research sector, makes use of 'lifestyle geocodes' to estimate income distributions down to postcode levels. These estimates generally calculate income distributions using a combination of market research records, postcode level geo-demographic indicators, national census and survey data and statistical imputation (Webber 2004). In effect, the national income distribution is used as the basis of the income distribution of individual postcodes, but its mean and standard deviation are allowed to vary between postcodes (Williamson and Voas 2000). However, the dependence on lifestyle surveys with potentially unknown response bias on a small number of lifestyle categories and on the imputation of household characteristics for postcode level address records has been criticised (Williamson and Voas 2000; McLoone 2002).

Finally, Birkin and Clarke (1988, 1989) introduced an approach that sought to combine elements of several of the above to produce not a modelled estimate but synthetic population microdata from which relevant aggregates and summaries could be calculated. This approach emphasised the importance of small-area income estimates and was the first study to use what they termed a spatial microsimulation method. This method used a combination of Monte Carlo sampling and iterative proportional fitting to produce small-area level microdata for each ward of the city of Leeds.

Since this original work, there have been considerable developments and advances in data availability and computing resources which have enabled experimentation with new techniques that can more easily and efficiently generate more reliable small-area microdata. In this context, Williamson et al. (1998) explored different solutions to finding the combination of UK Census Household SARs which best fit known small-area constraints. They tested various techniques of probabilistic combinatorial optimisation methods such as hill climbing algorithms, simulated annealing approaches and genetic algorithms in order to reweight cases in the SARs so that a good fit to known census-derived data was achieved when the estimations were re-aggregated to the small-area level. Building on these approaches, Ballas et al. (1999) report the testing of a number of approaches including the deterministic iterative proportional fitting method. Ballas then applied this method to the estimation of small-area level trends in equivalised income in York and Leeds between 1991 and 2001 using a combination of census and British Household Panel Survey (BHPS) data (Ballas 2004). He concluded that the iterative proportional fitting method was preferable on a number of dimensions including its deterministic nature and relatively efficient algorithm.

More recent work has sought to improve on these initial approaches through the further refinement of methods of error estimation (Smith et al. 2009) and of the selection of small-area constraints (Chin and Harding 2006; Anderson 2007; Tanton et al. 2009a, b; Anderson et al. 2009b).

In summary, a range of approaches to the small-area estimation of income have been developed. However, very few have attempted to produce measures of the proportion of households below a given income threshold which are now considered standard in the analysis of income inequalities. In this context, the remainder of this chapter presents a method for estimating the percentage of households whose net equivalised income is below 60% of the national median. The method uses the deterministic iterative proportional fitting approach to produce population microdata for each Welsh LSOA using a large sample survey and Census 2001 data for Wales.

4.3 The Iterative Proportion Fitting Approach

As we have seen in the last section, there has been considerable progress in the use of spatial microsimulation to produce small-area estimates. The reweighting methodologies briefly introduced above offer considerable potential for the creation of synthetic small-area microdata through the reweighting of national or regional survey microdata, such as the Family Resources Survey (FRS), using data from the census of population. Put simply, the method allocates all households from the sample survey to each small area and then, for each small area, reweights each household so that the derived small-area level tables of aggregate statistics for those reweighted households match identical tables from the UK Census 2001 (Williamson et al. 1998). This reweighting requires the identification of suitable constraint variables that must exist in both the small-area (census) and survey data in identical form (Williamson et al. 1998 and see Chap. 2 for the importance of preparing the raw data). It is these constraints that are the subject of the reweighting process.

As Williamson et al. (1998) and Williamson and Voas (2000) point out, there are many ways in which reweighting methodologies can be fine-tuned through the evaluation of the use of more or different census small-area tables or by changing the model parameters. These design choices can be summarised as:

- Choice of reweighting algorithm
- Choice of constraints to be used in reweighting
- Selection of households from the survey to be used for each small area
- Whether or not to require integer weights (i.e. produce 'whole households')

A wide range of techniques have been proposed for the reweighting of cases ranging from iterative proportional fitting through simulated annealing to linear programming and complex combinatorial optimization and generalised regression methods (Williamson et al. 1998; Ballas et al. 1999; Ballas and Clarke 2001; Tanton et al. 2011). Birkin and Clarke (1988, 1989) demonstrate how iterative proportional fitting (IPF) and Monte Carlo sampling can be employed to generate a wide range of attributes at the small-area level. The IPF method is well established and appears in a multitude of guises, from balancing factors in spatial interaction modelling through the RAS method in economic accounting, and its behaviour is relatively well known (Birkin and Clarke 1988; Wong 1992; Simpson and Tranmer 2005).

In essence, the method we have developed allocated all (or a specific selection of) households from the FRS to each Welsh LSOA and then iteratively reweighted each case using the iterative proportional fitting algorithm so that LSOA level tables of aggregate statistics matched identical tables from the UK Census 2001.

4.3.1 Definition of Income

The income survey data used was the Welsh subsample of the FRS 2003/2004 and FRS 2004/2005, and the income variable used was the sum of all net household incomes from:

- Earnings and self-employment (net of income tax and national insurance payments)
- Investments
- Disability benefits
- Retirement pensions plus any income support or pension credit
- Working tax credit and/or child tax credit received
- Other pensions
- Other benefits
- Other/remaining sources

In order to align the income values with the official UK Department for Work and Pensions' Households Below Average Income (HBAI) definitions (DWP 2007: Appendix 1), the following expenditures were then removed to produce the net income before housing costs (BHC):

- Domestic rates/council tax
- Contributions to occupational pension schemes (including all additional voluntary contributions (AVCs) to occupational pension schemes and any contributions to stakeholder and personal pensions)
- Insurance premiums made in case of sudden loss of earnings
- All maintenance and child support payments, which are deducted from the income of the person making the payment
- Parental contributions to students living away from home
- Student loan repayments

To calculate after housing costs (AHC) income, 'the total amount spent on water and sewerage rates, rent, mortgage interest, household rent, structural insurance (adjusted for combined cases to be consistent with HBAI) and service charges' (DWP 2007: Appendix 1) was removed from the before housing costs income variable.

In common with the official HBAI definition, the UK Department of Work and Pensions' variation of the Organisation for Economic Co-operation and Development's modified equivalisation scale was[4] then used to control for household composition and to produce an equivalised measure of household income before and after housing costs. These were then used as the basis for the calculations of the Welsh BHC/AHC medians and thence the allocation of households to the two

[4] Modified OECD scale $= 1 + 0.5 \times$ number of adults $+ 0.2 \times$ number of dependent children < 14; HBAI scale (BHC) $= 0.67 \times 1$ adult $+ 0.33 \times$ number of further adults $+ 0.2 \times$ number of children aged < 14; HBAI scale (AHC) $= 0.58 \times 1$ adult $+ 0.42 \times$ number of further adults $+ 0.2 \times$ number of children aged < 14.

Table 4.1 FRS BHC/AHC households – below average income results for Wales

		BHC	AHC
2003/2004			
	N	1,278	1,278
	% HBAI	12.51%	17.76%
2004/2005			
	N	1,239	1,239
	% HBAI	13.48%	16.87%
2003/2005			
	N	2,517	2,517
	% HBAI	12.99%	17.32%

indicator groups – above or below 60% of the relevant Welsh median to create a within-Wales poverty indicator (Table 4.1).

It should be noted that households with negative incomes were retained. Households reporting negative BHC income constitute 0.55% of Welsh households in 2003–2004 (0.81% in 2004/2005), whilst 0.86% (1.37%) report negative AHC income. It was not expected that retaining households with negative incomes will have any significant effect on the indicators as they will not substantially affect the median calculations.

4.3.2 Choice of Constraint Variables

Having determined to use IPF to reweight the survey observations to fit the small-area tables, it was then necessary to identify the constraint variables on which the IPF process would operate. The set of constraint variables must be:

1. Common to both the FRS and the census or at least derived from them
2. Available at the household level – as the poverty indicator to be estimated is at the household level
3. Known to be reasonable predictors of the indicator at the small-area level
4. Reasonably good predictors of the indicator at the micro (household) level

A review of census and FRS data was used to produce a list of variables that satisfied criteria 1 and 2, and recommendations from the literature (Williamson and Voas 2000; Williamson 2005) were used to filter these variables according to criteria 3 to produce a list of candidate constraints (Anderson 2009).

Finally, criteria 4 was tested within the FRS using standard logistic regression techniques to model the relationship between the micro-level constraints and the probability of a household having a net equivalised income below 60% of the national median. The *r*-squared value was used as an indicator of the value of the constraint variables, but in contrast to previous work which reported the use of repeated bivariate regressions to test each variable independently (Chin and Harding 2006), a stepwise multivariate method was used. The multivariate approach meant that correlations between constraint variables were taken into account and thus the 'pure' effects of each constraint were revealed whilst the use of the stepwise technique automatically included only those variables which had a statistically significant effect on the model and ordered the resulting indicators in decreasing order of their effects.

The overall model pseudo r-squared score can then be used as an indicator of how well the included constraints predict the outcome variables (in this case the BHC or AHC HBAI) at the household level, and is thus some indicator of the confidence we can have in the robustness of the eventual results. In addition, because the IPF technique iteratively reweights a series of constraints, the last constraint is necessarily fitted perfectly. It is therefore important that the constraints are used in an order that represents their increasing predictive power so that the 'best' constraint is fitted last and the stepwise results allowed us to establish this ordering.

Table 4.2 summarises these results and shows that we can be justified in pooling the 2003–2004 and 2004–2005 FRS data since the predictors of each indicator at the household level were essentially identical, although it is interesting to note that with the larger pooled sample (2003–2005), there were additional significant constraint variables: Household Response Person (HRP) gender in the case of BHC and HRP age in the case of AHC.

4.3.3 Small-Area IPF Algorithm Implementation

As previously discussed, these constraints were then used at the small-area (LSOA) level to iteratively reweight the FRS to fit each Welsh LSOA and so produce an estimate of the % HBAI for each LSOA for each indicator. Whilst results for 2003–2004 and 2004–2005 were generated separately, we report only those for the pooled 2003–2005 data using the constraints identified above.

Following Ballas et al. (2005a), we implemented a *regional weighting* scheme so that only households belonging to the same region as the particular LSOA are allocated to it. Ballas et al. also report using a process of integerisation to select the 'best fit' n weighted households for a given area where *n* was the number of households required for the ward. This integerisation process assigns integer weights to each household in the survey. Ballas et al. report that this integerisation produced some extremely poor results when tested against the census distributions and described a swapping algorithm to swap households between their 1991 wards in order to reduce errors and produce a better fit.

Since it is inevitable that the integerisation process would reduce within-zone variation, and for our purposes it was not necessary that each small area was allocated a whole number of households, we did not implement the integerisation process. Instead, our simplified method allowed the final household weights for each small area to remain fractional so that all possible survey households were retained. In our experience, this simplified method produced distributions that performed at least as well as Ballas et al.'s more complex combination of integerisation and household swapping.

Table 4.2 Significant constraints (in decreasing order of explanatory power)[a]

	2003–2004		2004–2005		2003–2005 pooled	
	BHC	AHC	BHC	AHC	BHC	AHC
	Employment status	Employment status	Employment status	Employment status	Employment status	Employment status
	Number of earners	Tenure	Number of earners	Number of earners	Number of earners	Tenure
	Tenure	Number of earners		Tenure	Tenure	Number of earners
					Gender of HRP	HRP age
Pseudo r-squared	0.122	0.234	0.126	0.219	0.129	0.232

[a]See Anderson (2009) for full results

Table 4.3 Small-area table for number of earners derived from Census 2001 for the first LSOA in Wales

Zone code	Number of households	Number of earners = 0	Number of earners = 1	Number of earners = 2	Number of earners = 3+
W01000001	517	294	132	85	6

Table 4.4 Small-area table for number of earners derived from the FRS 2003/4/5 for Wales

Number of households	Number of earners = 0	Number of earners = 1	Number of earners = 2	Number of earners = 3+
1,308	608	333	320	47

Table 4.5 First four zone 1 households with initial weights

Case	Region	HRP age	Number of rooms	Number of persons	NS-SEC of HRP	Composition	Number of earners	w_i
26,115	10	2	3	0	1	2	1	1
26,116	10	2	2	0	3	2	0	1
26,117	10	2	3	4	0	0	2	1
26,118	10	4	3	0	0	2	1	1
..	..	..	..	..	..	..	..	1

The objective was to produce a set of weights linking all households from the relevant government office region to all LSOAs in that region in the sense that the weights represent the 'fractional existence' of each household in each LSOA. Conceptually, the results can be thought of as a matrix of LSOAs (rows) and households (columns) where each cell contains the weight for a given household in a given LSOA.

To do this, two sets of tables were required for each constraint for each LSOA: the Census 2001 small-area tables for the constraints (see Table 4.3) and the analogous small-area tables constructed from the FRS households for the region in which the zone was found (see Table 4.4).

Starting with LSOA 1, all household weights (w_i) were initially set to 1 (see Table 4.5), whilst the weights for households that did not belong to the same region as the area in question were set to 0 rather than w_i to implement the regional weighting scheme.

Then, for each constraint in turn, the weights were adjusted using the formula:

$$N\mathrm{w}_h = w_{\mathrm{ih}} * \mathrm{c}_{\mathrm{hj}} / \mathrm{s}_{\mathrm{hj}}$$

where $N\mathrm{w}_h$ was the new household weight for household h, w_{ih} was the initial weight for household h, c_{hj} was element hj of the census data table (Table 4.3) and s_{hj} was element hj of the FRS statistical table (Table 4.4).

As an example, using the number of earners constraint, Table 4.6 shows the calculations for the first weights for the first four households so that the FRS sample fits the census distributions on this one dimension.

Table 4.6 First four zone 1 households with weights after fitting to constraint 1

Case	Region	Number of earners	W_1
26,115	10	1	$= 1 \times (132/333) = 0.396$
26,116	10	0	$= 1 \times (294/608) = 0.484$
26,117	10	2	$= 1 \times (85/320) = 0.266$
26,118	10	1	$= 1 \times (132/333) = 0.396$
..	..	..	..

Having adjusted the weights for the first constraint, the process then moves sequentially through each constraint variable, multiplying each new weight by that produced by the previous step. Since the last constraint to be fitted will necessarily be fitted perfectly, it was necessary to order the variables in order of the contribution to the *r*-squared in the regression model fitted (Table 4.2). This means the last variable to be fitted was the one which accounted for the most variation in the outcome variable of interest (% HBAI in this case).

Having passed over all constraints once, the process then looped back to the first constraint variable and repeated the reweighting starting from the weight produced in the last step (by the last constraint). Ballas et al. (2005a) found that iterating the procedure between five and ten times produced weights that reduced the error in fitting households to areas to a point where it no longer declined. Our experimentation suggested that ten iterations were sufficient to achieve a stable indicator value (Anderson 2009). Thus, after iterating the reweighting procedure ten times, the simulation then moved on to the next zone and repeated the process.

Calculating the % HBAI was thus a straightforward matter of summing the weighted indicator for each area and dividing by the number of households in that area.

4.4 Results

Figure 4.1 compares estimated before housing costs and after housing costs % HBAI for LSOAs in Wales. It can be seen that the two indicators produced slightly different results at the upper end of the distribution (higher income deprivation), although rather similar results for the least deprived. This was confirmed by a Spearman rank correlation of only 0.679 and by Fig. 4.1.

The spatial distributions of the BHC indicator (not shown) suggested a higher concentration of poor households in the former mining areas of South Wales and in the coastal areas on the Pembrokeshire/Ceredigion border as well as in other pockets in specific urban areas. In contrast, the areas with the highest % HBAI according to the AHC indicator (Fig. 4.2) were concentrated in the Valleys and South Wales urban areas.

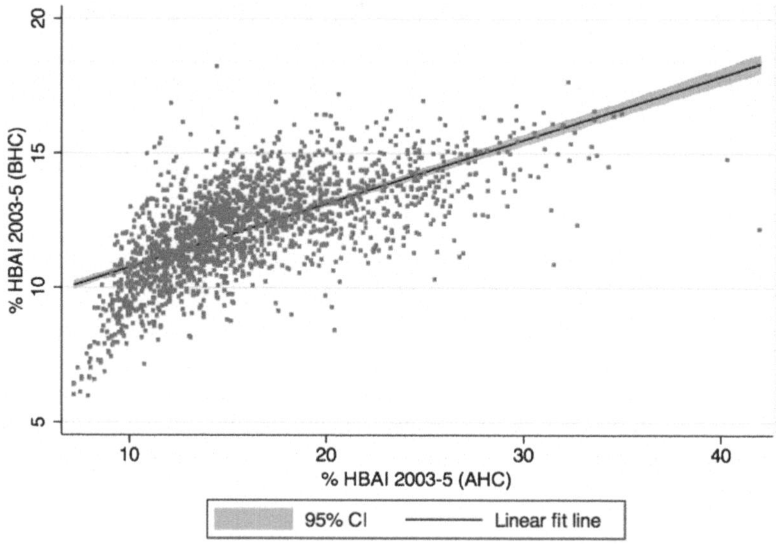

Fig. 4.1 Comparison of BHC and AHC HBAI indicators at LSOA for Wales (Census 2001, FRS 2003–2005 pooled)

Overall, the AHC indicator was considerably more diffuse in its distribution and thus may be a better 'relative poverty' indicator in comparison to the rather tighter BHC distribution which supports less differentiation (Fig. 4.3).

It should therefore be apparent that the utility of each indicator in a revised index of multiple deprivation will depend on the political objective and outcome desired since they reveal slightly different patterns of poverty.

4.5 Validation

In order to test the validity of the estimated distributions of the % HBAI, we made three kinds of comparisons:

- Comparison of the modelled results with reliable survey results at regional or country level to check internal validity and that the process accurately recreated inter-regional or inter-country variation. In this case, we used the FRS.
- Comparison of the estimated constraint counts with initial census constraint counts to check internal validity. This was the analysis of total absolute error (TAE) discussed in Ballas et al. (1999, 2006) and Smith et al. (2009).
- Compare estimated LSOA level results with other known small-area estimates. In this case, no equivalent small-area data were available, but instead, we compared the results to the income domain score of the Welsh IMD 2005.

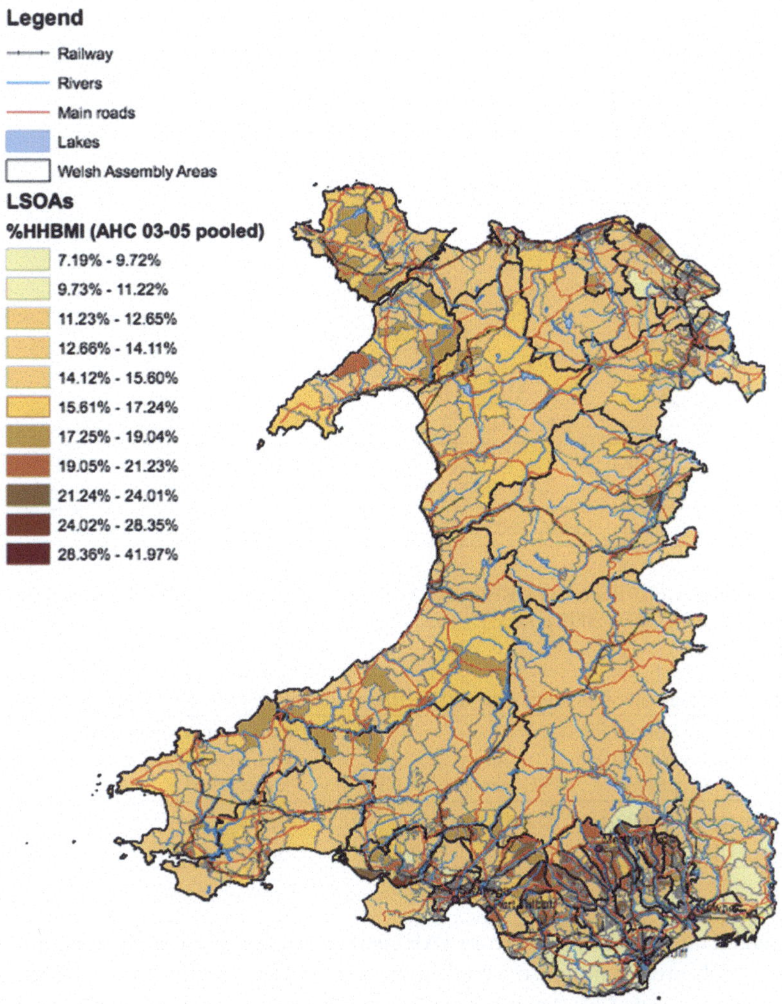

Fig. 4.2 Spatial distribution of % HBAI in Wales (AHC, LSOA level, Census 2001, FRS 2003–2005 pooled, natural breaks (Jenks))

Table 4.7 shows the % HBAI indicators (and 95% confidence interval) as calculated from the relevant FRS data and estimated from the spatial microsimulation process. Overall, there appeared to be a tendency to slightly underestimate % HBAI compared to the FRS results. In general, we would expect the microsimulation result to lie within the 95% confidence interval of the survey estimate, and as can be seen, the spatially microsimulated estimates provided a reasonable fit since they lay within these boundaries.

Turning to the constraint count error analysis, by entering the constraint counts as variables to be estimated, it is possible to compare the initial 'true' census constraint household counts with the estimated counts following the spatial microsimulation

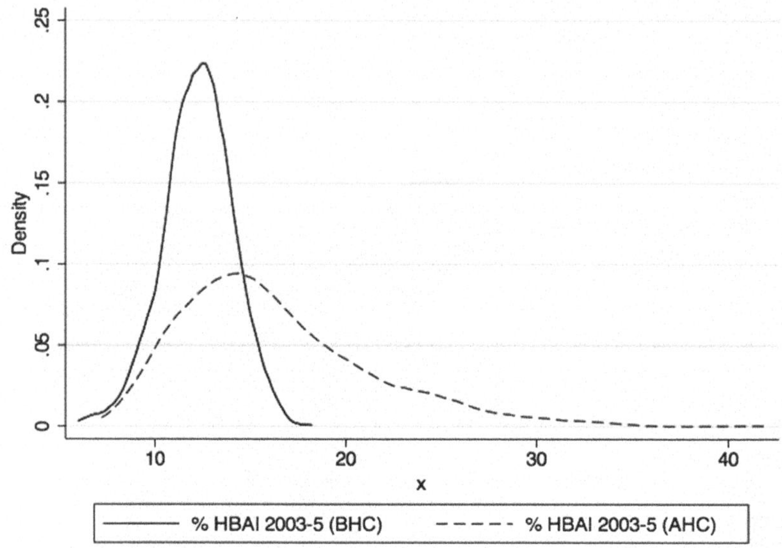

Fig. 4.3 Kdensity distributions of BHC and AHC % HBAI indicators at LSOA level for Wales (Census 2001, FRS 2003–2005 pooled)

Table 4.7 Comparison of simulated mean % HBAI results with FRS 2003–2005 pooled results

	Source FRS 2003–2005 pooled	SE mean	95% CI (+/–)	Spatial microsimulation (FRS 2003–2005 pooled/ Census 2001)	Difference
% HBAI (BHC)	12.992%	0.670	1.314	12.259%	–0.733%
% HBAI (AHC)	17.322%	0.754	1.479	16.294%	–1.028%

procedure. The total absolute error (TAE) is the difference for each constraint category for each area summed over all areas, whilst the standardised absolute error (SAE) is TAE divided by the number of units (in this case households).

Whilst minimising the difference between the 'true' and estimated counts is the objective, it is not yet clear in the literature what values of error are acceptable, although Smith et al. suggest that an SAE of less than 20% and ideally less than 10% in 90% of the areas is desirable, especially where the prevalence rate of the phenomenon of interest is low (Smith et al. 2009).

The % HBAI models for Wales performed substantially better than this, and elsewhere, we disaggregated the SAE to reveal the constraints that showed the poorest fit (Anderson 2009). The analysis suggested that levels of error were relatively low for both indicators with the largest error being for households with no earners (11.9% in each case). The mean error was no larger than 2.1% for any constraint, and in all cases, 90% of areas had SAE rates of less than 4%. As expected, we also confirmed that the order of the constraints means that the last category to be fitted (HRP Employment) fits perfectly.

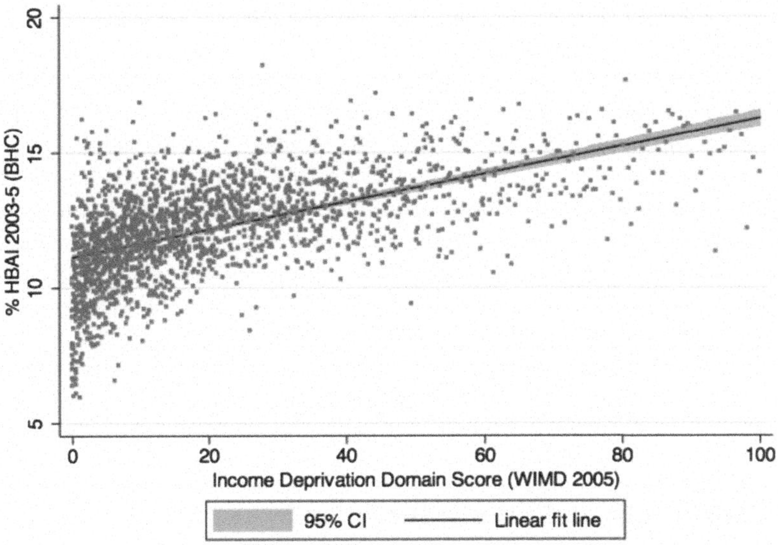

Fig. 4.4 WIMD 2005 income domain score vs spatial microsimulation results for BHC indicator (Spearman rho = 0.6041)

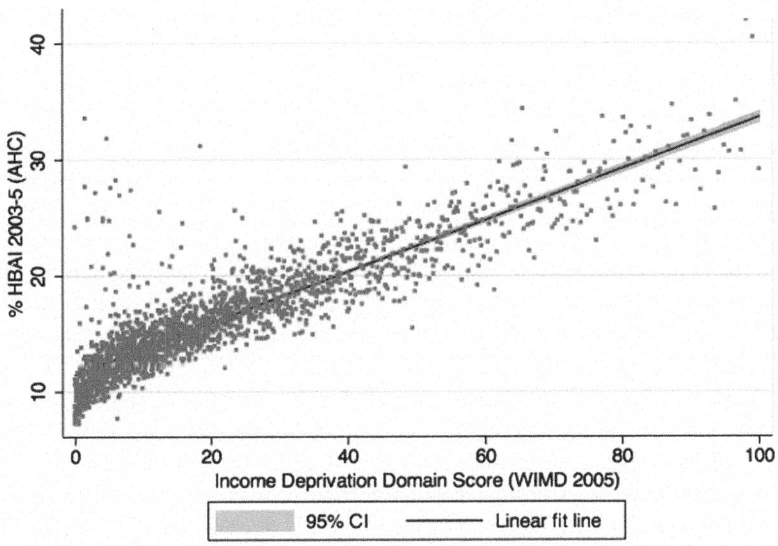

Fig. 4.5 WIMD 2005 income domain score vs spatial microsimulation results for AHC indicator (Spearman rho = 0.8834)

Finally, Figs. 4.4 and 4.5 show the fit between the Welsh IMD 2005 income domain score and the simulated % HBAI indicators at LSOA level. As expected, there was a strong rank order correlation between % HBAI using the equivalised indicator and the WIMD income domain score, and this was especially the case for the after housing costs indicator.

Figure 4.4 suggests the presence of a number of outliers which were low on the WIMD 2005 income score but relatively high on the simulated % HBAI. This was particularly noticeable in the case of the AHC indicator. Deeper analysis revealed that the two LSOAs which were in both the top 10% of HBAI (AHC) and the bottom 10% of the WIMD 2005 income score[5] had a high or relatively high proportion of students (who could not claim relevant benefits) according to the 2001 census. This suggested that one of the main differences between the Indices of Deprivation income domain results and the HBAI (AHC) results was the inclusion of low-income student households. LSOAs with higher proportions of students were therefore likely to appear to be 'more deprived' using the HBAI indicator than would be the case for the WIMD income domain score.

4.6 Conclusions and Future Directions

Overall, the results of this preliminary work on estimating the proportion of households below HBAI at the small-area level were encouraging. The results provided a synthetic household dataset which reproduced the Welsh %HBAI (BHC/AHC) as measured by the UK Family Resources Survey and which also produced a good fit to the Welsh IMD 2005 income domain score at the small-area level. This was especially true for the AHC measure.

The results also suggested that a focus on %HBAI, and especially on the AHC indicator, would present slightly different spatial distributions of income deprivation than would the WIMD 2005 income domain score. In particular, there would be differences where students make up a high proportion of survey respondents. This is of course likely to affect urban rather than rural areas. In addition, the differences between the results for the BHC and AHC indicators mean that consideration needs to be given to which would be the 'best' one to use in a future revision of the Welsh IMD. This cannot be answered by this chapter as it is dependent on the policy context and the uses to which the index and its components will be put.

The analysis of errors (SAE) suggested that in some small areas, the spatial microsimulation method produced a less accurate estimate than in others. This may have been because these areas were made up of an unusual combination of household types and future work could investigate extending the spatial microsimulation method to account for such areas and thus to reduce overall error still further.

The iterative proportional fitting method itself is performed in a robust, deterministic manner in this context. This determinism meant that variations in input data coding, constraint ordering or small-area table recoding were the only source of

[5] Somewhat counter intuitively, this means their WIMD 2005 income score would be low (i.e. not deprived).

variation in the small-area estimates. This proved extremely useful because it allowed the testing of different combinations of constraints and data coding options without the additional uncertainty caused by a probabilistic reweighting method.

As we have discussed, we also assumed that the order of the reweighting iterations mattered since the last constraint always fitted perfectly. We assumed that this then necessitated the use of the nested regression analysis to determine the 'best' constraint variable order. However, there is as yet no evidence that this 'best' order produces a substantially better fit than, for example, a random ordering, and we anticipate testing this assumption in future work.

Finally, we would also suggest that the ability of the IPF algorithm to produce fractional weights proved crucial to the reconstruction of accurate aggregated estimates for comparison with the original survey data (Table 4.7). It also enabled us to retain all relevant households in our synthetic small-area samples and so increase their (weighted) heterogeneity. This would prove crucial if these data were then to be used as a basis for the microsimulation modelling and thus small-area analysis of, for example, policy intervention scenarios.

Acknowledgements This work is based on data provided through EDINA UKBORDERS with the support of the ESRC and JISC and uses boundary material which is copyright of the crown and the post office.

Census data was originally created and funded by the Office for National Statistics and was distributed by the Census Dissemination Unit, MIMAS (University of Manchester). Census output is crown copyright and is reproduced with the permission of the controller of HMSO.

The FRS is collected and sponsored by the Department for Work and Pensions and is distributed by the UK Data Archive, University of Essex, Colchester. FRS data is copyright and is reproduced with the permission of the Controller of HMSO and the Queen's Printer for Scotland.

The WIMD 2004 was constructed by the Social Disadvantage Research Centre at the Department of Social Policy and Social Research at the University of Oxford and distributed by the Welsh Assembly Government.

We would also like to thank Professor Holly Sutherland (ISER, University of Essex) for advice on the treatment of negative income and housing costs.

This work was sponsored by the Welsh Assembly Government.

References

Anderson, B. (2007, July). *Cash in, cash out: Spatially microsimulating household income and expenditure at small area levels*. Paper presented at the Royal Statistical Society Conference 2007, University of York, York, UK.

Anderson, B. (2008). *Time to play: Combining time-use surveys and census data to estimate small area distributions of potentially ICT mediated leisure*. Paper presented at the AoIR 8, October 17, 2007, Simon Fraser University, Vancouver, BC, Canada.

Anderson, B. (2009). *Welsh small area estimates of income deprivation*. Colchester: University of Essex.

Anderson, B., De Agostini, P., Laidoudi, S., Weston, A., & Zong, P. (2009a). Time and money in space: Estimating household expenditure and time use at the small area level in Great Britain. In A. Zaidi, A. Harding, & P. Williamson (Eds.), *New frontiers in microsimulation modelling: Public policy and social welfare Vol. 36*. Aldershot: Ashgate.

Anderson, B., De Agostini, P., & Lawson, T. (2009b, July). *Estimating income, expenditure and time-use within small areas*. Paper presented at the ESRC Microsimulation Seminar Series Workshop III 'Moving beyond tax-benefit and demographic modelling', University of Leeds, Leeds, UK.

Ballas, D. (2004). Simulating trends in poverty and income inequality on the basis of 1991 and 2001 Census data: A tale of two cities. *Area, 36*(2), 146–163.

Ballas, D., & Clarke, G. (2001). Modelling the local impacts of national social policies: A spatial microsimulation approach. *Environment and Planning C: Government and Policy, 19*, 587–606.

Ballas, D., Clarke, G., & Turton, I. (1999, July). *Exploring microsimulation methodologies for the estimation of household attributes*. Paper presented at the 4th International Conference on GeoComputation, Mary Washington College, Fredericksburg, VA.

Ballas, D., Clarke, G., Dorling, D., Eyre, H., Thomas, B., & Rossiter, D. (2005a). SimBritain: A spatial microsimulation approach to population dynamics. *Population, Space and Place, 11*, 13–34.

Ballas, D., Clarke, G., Dorling, D., Rigby, J., & Wheeler, B. (2005b, September). *Using geographical information systems and spatial microsimulation for the analysis of health inequalities*. Paper presented at the 10th International Symposium on Health Information Management Research – iSHIMR 2005, CITY Liberal Studies, Thessaloniki, Greece.

Ballas, D., Dorling, D., Anderson, B., & Stoneman, P. (2006). *Assessing the feasibility of producing small area income estimates: Phase I project report*. Sheffield: Department of Geography, University of Sheffield.

Bates, A. (2006). Methodology used for producing ONS's small area population estimates. *Population Trends, 125*, 30–36.

Birkin, M., & Clarke, M. (1988). SYNTHESIS – A synthetic spatial information system for urban and regional analysis: Methods and examples. *Environment and Planning A, 20*, 1645–1671.

Birkin, M., & Clarke, G. (1989). The generation of individual and household incomes at the small area level using synthesis. *Regional Studies, 23*(6), 535–548.

Birkin, M., & Clarke, M. (2011). Spatial microsimulation models: A review and a glimpse into the future. In J. Stillwell & M. Clarke (Eds.), *Population dynamics and projection methods*. London: Springer.

Chin, S. F., & Harding, A. (2006). *Regional dimensions: Creating synthetic small-area microdata and spatial microsimulation models* (Technical Paper 33). National Centre for Social and Economic Modelling, Canberra: University of Canberra.

Druckman, A., & Jackson, T. (2008). Household energy consumption in the UK: A highly geographically and socio-economically disaggregated model. *Energy Policy, 36*(8), 3177–3192.

DWP. (2007). *Households below average income (HBAI) 1994/95–2005/06*. London: Department of Work and Pensions.

Edwards, K. L., & Clarke, G. P. (2009). The design and validation of a spatial microsimulation model of obesogenic environments for children in Leeds, UK: SimObesity. *Social Science & Medicine, 69*(7), 1127–1134.

Eurostat. (2007). Inequality of income distribution (S80/S20 income quintile share ratio), at-risk-of-poverty rate and at-persistent-risk-of-poverty rate. *Key indicators on EU policy – Structural indicators – Social Cohesion – Living conditions*. Retrieved July 4, 2008, from http://europa.eu.int/estatref/info/sdds/en/strind/socohe_di_base.htm

Gong, C., McNamara, J., Vidyattama, Y., Miranti, R., Tanton, R., Harding, A., & Kendig, H. (2011). Developing spatial microsimulation estimates of small area advantage and disadvantage among older Australians. *Population, Space and Place*. doi:10.1002/psp. 692.

Gordon, D., & Townsend, P. (2000). *Breadline Europe: The measurement of poverty*. Bristol: Policy Press.

Gosh, M., & Rao, J. K. (1994). Small area estimation: An appraisal. *Statistical Science, 9*(1), 55–76.

Harding, A., Vidyattama, Y., & Tanton, R. (2011). Demographic change and the needs-based planning of government services: Projecting small area populations using spatial microsimulation. *The Journal of Population Research, 28*(2–3), 203–224.

Heady, P., Clarke, P., Brown, G., Ellis, K., Heasman, D., Hennell, S., et al. (2003). *Model-based small area estimation Series No. 2*. London: Office for National Statistics.

Marsh, C. (1993). Privacy, confidentiality and anonymity in the 1991 Census. In A. Dale & C. Marsh (Eds.), *The 1991 census user's guide* (pp. 111–128). London: Her Majesty's Stationary Office.

McLoone, P. (2002). *Commercial income data: Associations with health and census measures* (Occasional paper No 7). Glasgow: MRC Social & Public Health Sciences Unit.

Mohana, J., Twigg, L., Barnard, S., & Jones, K. (2005). Social capital, geography and health: A small-area analysis for England. *Social Science & Medicine, 60*, 1267–1283.

Morrissey, K., Clarke, G., Ballas, D., Hynes, S., & O'Donoghue, C. (2008). Examining access to GP services in rural Ireland using microsimulation analysis. *Area, 40*(3), 354–364.

Noble, M., Wright, G., Dibben, C., Smith, G., McLennan, D., Anttila, C., et al. (2004). *Indices of deprivation 2004*. London: Office of the Deputy Prime Minister.

Noble, M., Wright, G., Smith, G., & Dibben, C. (2006). Measuring multiple deprivation at the small-area level. *Environment and Planning A, 38*(1), 169–185.

Noble, M., McLennan, D., Wilkinson, K., Whitworth, A., Barnes, H., & Dibben, C. (2008). *Indices of deprivation 2007*. London: Communities and Local Government.

Rao, J. K. (2003). *Small area estimation*. London: Wiley.

Simpson, L., & Tranmer, M. (2005). Combining sample and census data in small area estimates: Iterative proportional fitting with standard software. *The Professional Geographer, 57*(2), 222–234.

Smith, D. M., Harland, K., & Clarke, G. (2007). *SimHealth: Estimating small area populations using deterministic spatial microsimulation in Leeds and Bradford*. Leeds: University of Leeds.

Smith, D., Clarke, G., & Harland, K. (2009). Improving the synthetic data generation process in spatial microsimulation models. *Environment and Planning A, 41*, 1251–1268.

Tanton, R., Mcnamara, J., Harding, A., & Morrison, T. (2009a). Small area poverty estimates for Australia's eastern seaboard in 2006. In A. Zaidi, A. Harding, & P. Williamson (Eds.), *New frontiers in microsimulation modelling: Public policy and social welfare Vol. 36*. Aldershot: Ashgate.

Tanton, R., Vidyattama, Y., McNamara, J., Vu, Q., & Harding, A. (2009b). Old, single and poor: Using microsimulation and microdata to analyse poverty and the impact of policy change among older Australians. *Economic Papers, 28*(2), 102–120.

Tanton, R., Vidyattama, Y., Nepal, B., & McNamara, J. (2011). Small area estimation using a reweighting algorithm. *Journal of the Royal Statistical Society: Series A (Statistics in Society), 174*(4), 931–951.

Vidyattama, Y., & Tanton, R. (2010). Projecting small area statistics with Australian spatial microsimulation model (SpatialMSM). *Australian Journal of Regional Studies, 16*(1), 99–126.

Vidyattama, Y., Cassells, R., Harding, A., & McNamara, J. (2011). Rich or poor in retirement? A small area analysis of Australian private superannuation savings in 2006 using spatial microsimulation. *Regional Studies*. doi:10.1080/00343404.2011.589829.

Webber, R. (2004). *The relative power of geodemographics vis a vis person and household level demographic variables as discriminators of consumer behaviour*. London: UCL.

Williamson, P. (2001). An applied microsimulation model: Exploring alternative domestic water consumption scenarios. In G. Clarke & M. Madden (Eds.), *Regional science in business*. London: Springer.

Williamson, P. (2005). *Income imputation for small areas*. Liverpool: University of Liverpool.

Williamson, P., & Voas, D. (2000). Income estimates for small areas: Lessons from the census rehearsal. *BURISA, 146*, 2–10.

Williamson, P., Birkin, M., & Rees, P. (1998). The estimation of population microdata by using data from small area statistics and samples of anonymised records. *Environment and Planning A, 30*, 785–816.

Wong, D. (1992). The reliability of using the iterative proportional fitting procedure. *Professional Geographer, 44*(3), 340–348.

Chapter 5
SimObesity: Combinatorial Optimisation (Deterministic) Model

Kimberley L. Edwards and Graham Clarke

5.1 Introduction

This chapter details a deterministic method of spatial microsimulation modelling, which uses a set of algorithms based on combinatorial optimisation. This model, called SimObesity, was developed within the School of Geography, University of Leeds. An application of this model to estimate adult obesity prevalence is demonstrated. The chapter discusses the value of adopting a spatial microsimulation procedure and briefly debates the pros and cons of probabilistic and deterministic techniques for data imputation. Having chosen the latter, the chapter discusses the data and methodology used to estimate small-area prevalence of obesity in northern England. The results are discussed both in terms of the reliability of the model outputs (validation) and in terms of the spatial variation in estimated patterns.

5.2 Why Use Spatial Microsimulation Modelling to Model Disease Data?

There are two key reasons for modelling disease data using a spatial microsimulation model. First, many chronic diseases are a result of multifaceted associations, including biological, physiological, environmental, social and economic factors. Understanding how all of these factors vary will facilitate an understanding of variations in the disease of interest.

K.L. Edwards (✉)
School of Clinical Sciences, University of Nottingham, Nottingham, UK
e-mail: Kimberley.edwards@nottingham.ac.uk

G. Clarke
School of Geography, University of Leeds, Leeds, UK

R. Tanton and K.L. Edwards (eds.), *Spatial Microsimulation: A Reference Guide for Users*, Understanding Population Trends and Processes 6, DOI 10.1007/978-94-007-4623-7_5,

Second, disease data are not always available at the required spatial scale for useful policy analysis. It is clear that different spatial scales draw attention to different disease associations (Gatrell 2002). Analysing data for large areas (e.g. a city like London or large area like Yorkshire) gives a broad perspective, but assumes homogeneity at this level and potentially falls foul of the ecological fallacy problem. Analysis for large areas can imply that spatial patterns are uniform over the region being studied and that all households in that region have the same health conditions. Individual conditions are thus not well modelled (Wilkinson et al. 1998).

Similarly, drilling down qualitatively to only a handful of individuals gives great perspective on those people's lives, but has limited public health perspective in terms of aggregating findings to the entire population. Thus, modelling at the small-area level (for instance, Census tract level) allows for greater heterogeneity across fairly small populations. Modelling data for small areas is also a cost-effective alternative to undertaking an expensive and time-consuming data collection in the study area.

5.2.1 *Why Use a Deterministic Model?*

There is a major choice to be undertaken at the start of the spatial microsimulation model building process: whether to use a probabilistic or deterministic model. Probabilistic models use random sampling at some point within the algorithm; deterministic models are rule based, that is, if A is true, then B occurs. The reader will note that the static spatial microsimulation models described in Chaps. 3, 4 and 7 are all probabilistic, whereas this model (and the one described in Chap. 6) are deterministic.

A probabilistic algorithm will produce a different result each time the model is run because effectively, the input files are changing as different random numbers are used. That is, if the model selects say 100 individuals, the selection is random and so the model simulation would be based on a different set of individuals each time the model is run. This issue can be resolved by running the model a number of times and using an average result. However, the key benefit of a deterministic approach is that the algorithm results in only one solution, that is, the same solution occurs each time the model is run (assuming the use of the same constraint variables and input datasets).

5.3 SimObesity Methodology

5.3.1 *Data Used for the Model*

The SimObesity methodology combines data from two sources. The first is a dataset aggregated at the required spatial scale including a variable descriptor for geography, such as postcode or output area. This will be referred to as the 'geographic dataset'. So in the UK, this would normally be the latest Census data available. The second

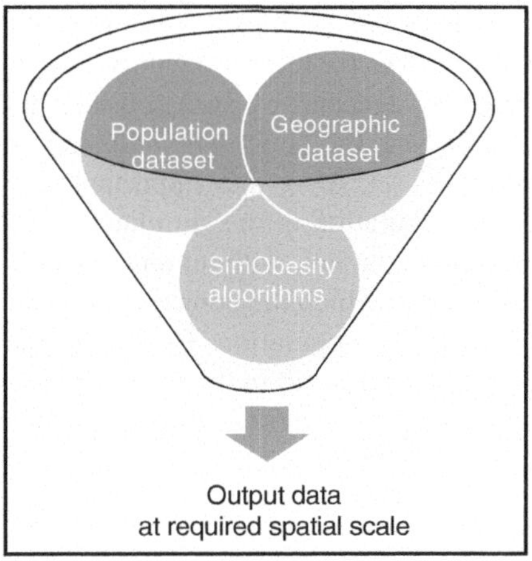

Fig. 5.1 Outline of SimObesity data combination

is a dataset at the individual level that contains the variable to be estimated. This will be referred to as the 'population dataset'. This second dataset does not need to have any geography variable descriptor, although this can be useful for validation. Often, these are datasets contained within nationally representative samples. Examples of suitable datasets in the UK would be the Health Survey for England, the British Household Panel Survey or the National Diet and Nutrition Survey. The choice of dataset will depend on the variable(s) being estimated, as the variable being estimated needs to be on the dataset. Further, a larger population dataset is preferred to a smaller one as it gives the algorithm more data to select from. A small dataset can lead to under- and overestimation at the extremes (e.g. for very deprived or very wealthy areas) (Huang and Williamson 2001; Birkin and Clarke 2011).

These two datasets are combined using the SimObesity model, and the resulting output dataset contains all the variables contained in the two starting datasets (see Fig. 5.1). The output dataset is an estimation of the population living in the study area for all of the variables in both starting datasets. Each individual in this dataset has been assigned to a micro-area (at the micro-spatial scale provided in the starting geographic dataset); thus, it is always possible to aggregate these data upwards for analysis.

As the two starting datasets are combined based on common variables between the two datasets, it is essential that the variables used to link the datasets are present in both datasets. These linkage variables are often called 'constraints' or 'benchmark' variables. For SimObesity, it is necessary to use constraints that are associated with the variable(s) being estimated (Edwards and Clarke 2009). These are the so-called optimising constraints. The simulation will only be as good as the underlying associations. Thus, the optimising constraints should be strong predictor(s) of the variable(s) being estimated.

Any variables that will be used in any subsequent analysis should also be included as constraints. For example, gender may be a variable we want to adjust for at some point in a study, so while it may not be associated with the variable being estimated, we would include it as a constraint. These are called 'elective constraints'.

SimObesity permits any number of constraints (whether optimising or elective). It also allows for them to be univariate (e.g. age) or to be cross-tabulated (e.g. marital status by age and by sex). Further, a constraint could be cross-tabulated in one dataset (say, the geographic dataset) but only available as a univariate table in the other dataset. Also, it is important to note that with this model, the resulting synthetic population in the simulated dataset is chosen based on all of the constraints used in the model, without any one constraint taking priority or preference. That is, the order in which the constraint variables are entered into the model does not affect the resulting population. Hence, a simulation using age, ethnicity, deprivation and gender would result in the same simulated population as one using gender, ethnicity, age and deprivation (assuming, of course, that the same two starting datasets are used for both models).

5.3.2 Estimation Methodology

SimObesity uses two algorithms. The first is a deterministic reweighting algorithm (to allocate people to the geographic areas); the second, an optimization algorithm (so only 'whole' people are selected).

The starting point is two datasets: the population dataset (with records for individuals but no geographic variable) and the geographic dataset (with aggregated data at the correct geographic level). Both datasets share some common variables, some of which will be used as constraint variables (whether optimising or elective constraints). These two datasets are subdivided into constraint and summary tables before being inputted into the model. So, for example, if six constraint variables are used, there would be six constraints and six summary tables. Each 'constraint' table is simply the total number of people for each category and area from the geographic dataset for that variable (rows: each area, columns: each categorization for the constraint variable). Each constraint table should therefore have the same total number of people residing in each area. If not, an adjustment should be applied to ensure the tables are consistent (Edwards et al. 2010a). The 'summary' tables are derived from the population dataset and are the total number of individuals for each category for each constraint variable (only one row; columns: each categorization for the constraint variable).

An initial weight is applied to each individual in the population dataset in order to compensate for bias/error. Then, for each constraint variable in turn, the reweighting algorithm (see Eqn. 5.1 and 5.2) determines what combination of individuals from the population dataset best matches the constraint aggregations for each geographic area. In Eqn. 5.1 and 5.2, P_{ij} denotes each individual in the population dataset, and X_{ij} denotes the weight for individual i in area j. On the first run through the equations (i.e. for the first constraint variable), W_{ij} is the individual's original weight in the population dataset. For subsequent constraint variables, it is the resulting weight (Y_{ij}) from the preceding constraint. C_{ij} denotes the constraint value for individual i in area j for each constraint table in turn. S_{ij} denotes the value for individual i in area j for each summary table in turn. ΣC_j is the sum of the corresponding area column for each constraint variable. ΣX_j is the sum of the corresponding area

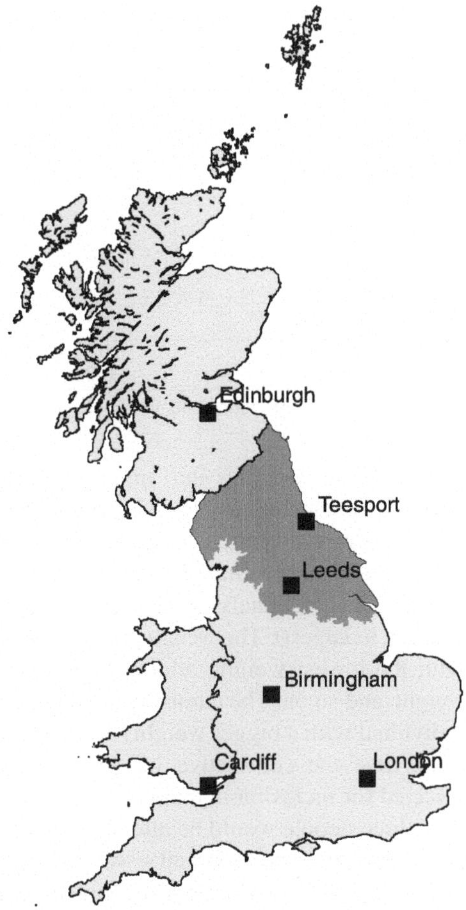

Fig. 5.2 Map of study area (north of England)

column for the reweight value determined in the preceding stage. A worked example of this is available elsewhere (Edwards and Clarke 2009).

For $P_{ij,}$

$$X_{ij} = W_{ij} \times C_{ij} / S_{ij}, \tag{5.1}$$

$$Y_{ij} = X_{ij} \times \sum C_j / \sum X_j. \tag{5.2}$$

The end result from this first algorithm is a new dataset containing a final weight for each individual for each area (rows: each individual from the population dataset, columns: each area), which symbolises the likelihood that an individual would reside in that area.

If the concluding weights are simply aggregated up, then the model allows for fractions of people, which is not a realistic prospect. Thus, the second algorithm seeks to only allocate whole people to the final dataset. That is, this second algorithm is an iterative optimisation methodology to convert the first algorithm's concluding

Table 5.1 Example of output from each of the two algorithms

Person ID	Weight resulting from the reweighting algorithm (algorithm 1)	Weight resulting from optimization algorithm (algorithm 2)
Person 1	0.0	0
Person 2	0.5	1
Person 3	2.3	2
Person 4	1.7	1
Person 5	3.2	4
Person 6	0.0	0
Person 7	0.2	0
Person 8	0.4	0
Person 9	1.7	2
Person 10	0.0	0

The right hand column represents the final synthetic population 'living' in this particular micro-area

weights from decimals into integers. Ballas et al. (2005) have tested the optimal integerisation techniques for this process and have suggested the following procedure (the SimObesity version of the algorithm uses a 'floor' function, which means that values are always rounded downwards).

First, the individuals are ranked in ascending order of weight for that area (i.e. from smallest to largest). The weights are aggregated one by one, starting with the individual with the lowest weight, adding the value for the individual with the next lowest weight, and so on. The cumulative weight continues to move down one row (i.e. to an individual with a bigger weight) at a time until each individual has been integerised. Each time the cumulative weight exceeds one, the corresponding individual is selected for inclusion in that area. Thus, if the new cumulative weight becomes 3.2, then three people would be allocated to that area and '0.2' carried forward to aggregate with the next individual's weight.

These data are then used to populate the study area. Using an example of a study area where only ten people live, the optimised weights given in Table 5.1 show that there are four synthetic versions of person 5 'living' in this area in the simulated dataset, two versions of both persons 3 and 9, but only one synthetic version of persons 2 and 5 and none of persons 1, 6, 7, 8 and 10. The full model produces these data for each area included in the model.

5.3.3 Validation Methodology

SimObesity does not automatically validate the estimates produced. It is therefore essential that the user spends some time both internally and externally validating their outputs.

It is recommended to start with internally validating the outputs, that is, to aggregate the estimated individual level data to the scale of the micro-areas in the starting geographic dataset and then to compare these results. It would be expected that there is no significant difference between the two datasets for all the constraint variables. Equal variance two-tailed *t* tests and linear regression analyses can be used to assess any differences between the estimated and actual datasets. Note, this internal validation is only applicable to the constraint variables (e.g. age, sex and deprivation), and not for the output variable (e.g. obesity). This is because the input datasets do not have output variable data at sufficient fine geographic scale to undertake these internal validation analyses (if they did, the spatial microsimulation of these data would not be required).

External validation is when data from another source on the variable being estimated are used to corroborate the validity of the estimates. For example, if the model were used to estimate how many obese people lived in a particular area, then a sample of obesity data would need to be collected from individuals actually living in that area, using the same measurement technique for obesity as used in the original population dataset (e.g. body mass index with a cut-off of 30 kg m^{-2}). This process is often more problematic than internal validation simply because data do not already exist at the micro-level, which is why the data are being estimated in the first place. However, it is sometimes possible to aggregate the estimated data to known data at a higher spatial scale (i.e. to match against known regional data). Although household or individual data on health conditions is rarely available, it is occasionally known. Smith et al. (2011) provide a very useful case study which, reassuringly, shows that the deterministic reweighting procedure can produce a very good fit against known data at the small-area level (when predicting small-area smoking rates in New Zealand).

5.3.4 SimObesity Application: Estimating Adult Obesity Prevalence

The next section of this chapter describes the process of estimating adult obesity data for a large part of northern England (see Fig. 5.2). It is universally accepted that obesity is a massive problem in the UK and worldwide, with serious medical repercussions for individuals. Yet obesity data are not routinely collected in the UK. Some sample data for the UK are available; the Health Survey for England 2008 showed that nearly a quarter of adults were obese and 65% overweight (Craig and Mindell 2008). However, these Health Survey for England data are not available at a fine spatial scale, and often health data have no geographical coding. Thus in order to scrutinise adult obesity data at the small-area level, it is necessary to estimate these data. Spatial microsimulation models are an ideal tool for this task, and there are many examples of their application to estimate health data (Tomintz et al. 2008; Procter et al. 2008; Edwards et al. 2010b; Morrissey et al. 2010).

5.3.4.1 Datasets

For this simulation, the population dataset used data from the Health Survey for England dataset (which is available from the Essex University archive website: http://www.data-archive.ac.uk/). This is because it contained individual level data on body mass index (BMI), which can be used to define obesity. BMI is a ratio of a person's weight and height, calculated as their weight in kilograms divided by the square of their height in metres. For adults, if this ratio is greater than 30 kg m^{-2}, the person is classified as obese. It also contained data on the required constraint variables (see below). To maximise the size of this dataset, 3 years worth of data were combined (2004, 2005 and 2006) (National Centre for Social Research and UCL 2004). This resulted in data on 36,525 individuals.

The geographic dataset was the Census 2001, at lower super output area (LSOA) level. An LSOA is a socially homogenous geographic unit in the UK based on Census data. Each LSOA has a minimum of 400 households and 1,000 residents, averaging 1,500 residents. The study area included 975 wards, which constitutes 4,318 Census LSOAs.

5.3.4.2 Model Constraint Variables

From all of the variables in common between the population and geographic datasets, age had the strongest correlation with BMI ($r=0.35$). Thus, it seemed sensible to include age as an optimising constraint variable. Age was provided as a continuous (integer) variable in the population dataset and categorised (0–15, 16–19, 20–49, 50–69, 70+ years) in the geographic dataset.

Sex and deprivation were selected as elective constraints because it was important that these variables were also accurately estimated. Sex was categorised as male or female in both the population and geographic datasets. Deprivation was calculated using the Index of Multiple Deprivation. This is a variable that is derived from a combination of various data within the 2001 Census. It combines many variables covering seven different domains, each of which is given a different weighting: income (22.5%), employment (22.5%), health/disability (13.5%), education (13.5%), housing (9.3%), crime (9.3%) and living environment (9.3%) (Communities and Local Government 2010). By amalgamating these variables, a single deprivation score is determined for each LSOA. The 2004 version was utilised rather than the current 2007 version (http://webarchive.nationalarchives.gov.uk/+/http://www.communities.gov.uk/archived/general-content/communities/indicesofdeprivation/216309/) because the oldest population dataset stemmed from 2004. The deprivation score was provided as a continuous variable in both the population and geographic datasets. For the simulation, it was categorised into quintiles: 0.6–8.3 (least deprived) and 8.4–13.7, 13.8–21.1, 21.2–34.2 and 34.3–86.4 (most deprived).

The geographic dataset was split into three constraint tables, and the aggregate constraint variable data were equalised using the mean total population across the three tables. This is because in the UK, the aggregate Census tables have slight imperfections introduced into them deliberately in order to protect individual confidentiality, resulting in slightly (1 or 2 people per thousand) different population totals for any one area depending on which variable table is being used. Similarly, the population dataset was aggregated into three summary tables. An initial weight of 1 was applied to each individual. The SimObesity model was then run, resulting in an individual level output (the number of each individual residing in each LSOA) and an aggregate level output (the number of individuals in each of the constraint and BMI categories for each LSOA).

5.3.4.3 Internal Validation

The output data were then internally validated, by comparing the estimated aggregate weighted figures for each constraint category for each ward to the actual aggregate figures. The ideal is that the estimated and actual figures are identical. Thus, to determine the accuracy of the model, an equal variance two-tailed *t* test for each constraint variable was used to determine whether there were any statistical differences between the aggregated weighted simulated data and the actual benchmark data from the Census. Additionally, in order to assess the precision of the model, univariate linear regression models were run to examine the relationship between the simulated and actual data for each category of the constraint variables (i.e. 12 models). For example, for each LSOA, the number of children aged 0–15 years in the simulated dataset can be compared with the actual number from the geographic dataset. An ideal, precise model would show that $y=x$, so a coefficient of determination is equal to 1.

The results of the internal validation of the simulated datasets show that the model estimates are robust. A scatter plot visualisation of the linear regression models is provided in Fig. 5.3. These clearly show how the data are clustered around the $y=x$ trend line, corroborated by the high coefficients of determination (see Table 5.2). Also, the equal variance two-tailed *t* tests showed that there were no significant differences between the simulated and actual datasets for any of the constraint variables (see Table 5.2).

5.3.4.4 External Validation

In order to externally validate this model, we compared the proportion of obese, and overweight/obese, people in the synthetic dataset with national figures for England. English data for 2004, 2005 and 2006 were amalgamated to coincide with the period of the simulated data population dataset and also because rates of obesity had been rising every year since 1997 (Craig and Mindell 2008). These results are presented in Table 5.3. There is less than 1% difference in the prevalence of obesity for adults, broken down as a 2% overestimate in female obesity and 2% underestimate in male obesity. While the difference in the prevalence of overweight and obese is slightly greater, it is still under 5% and due to a larger underestimate in the data for males. Of course, these differences could also be caused by regional variations, that is, geographical factors not accounted for in the model. Birkin and Clarke (2011) discuss the impacts geography can make on actual spatial variations as opposed to simulation results based only on socio-economic factors.

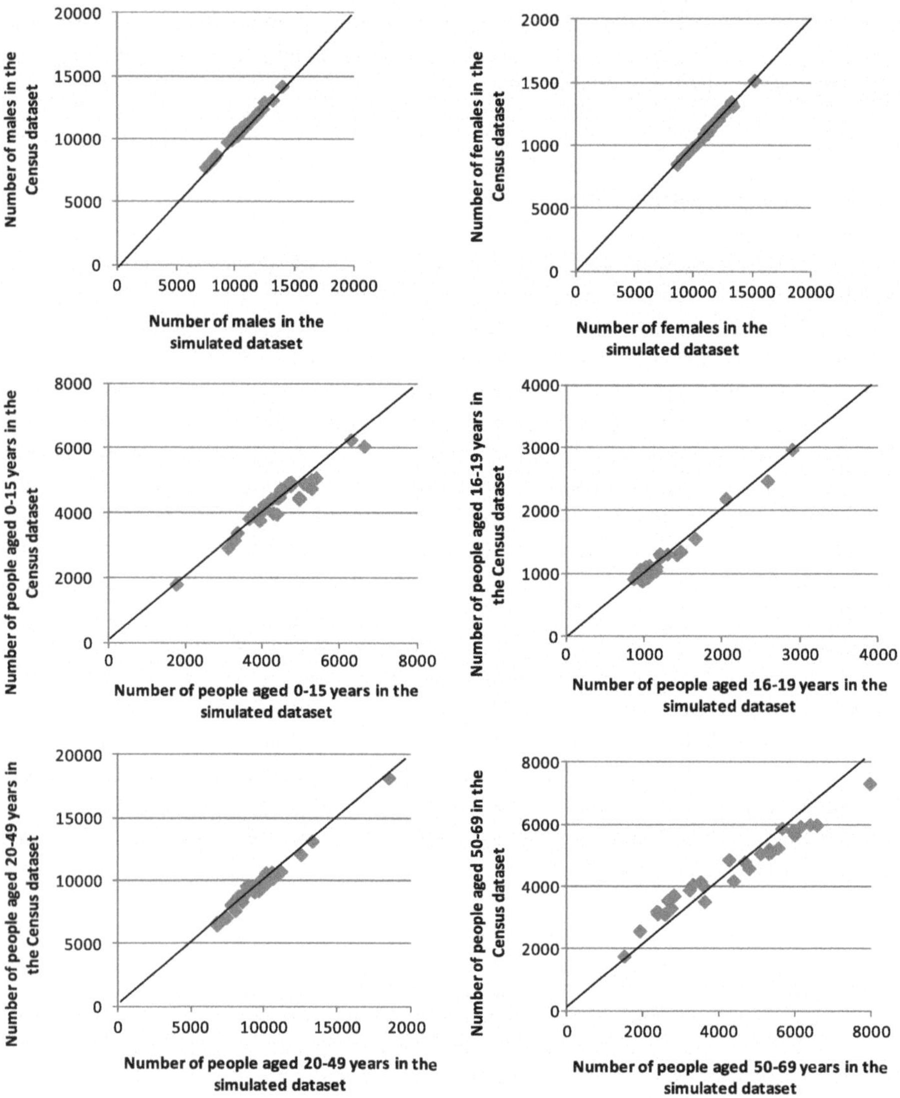

Fig. 5.3 Internal validation results: linear regression models for each of the three constraint variables (sex, age, deprivation). The line on each plot represents the 'dummy' $y=x$ line

5.4 Analyses of Obesity Data

The simulation estimated the population for the north of England, together with all their attributes from both datasets (i.e. both the HSE and Census). These data showed that 25% of residents were obese and 62% overweight or obese.

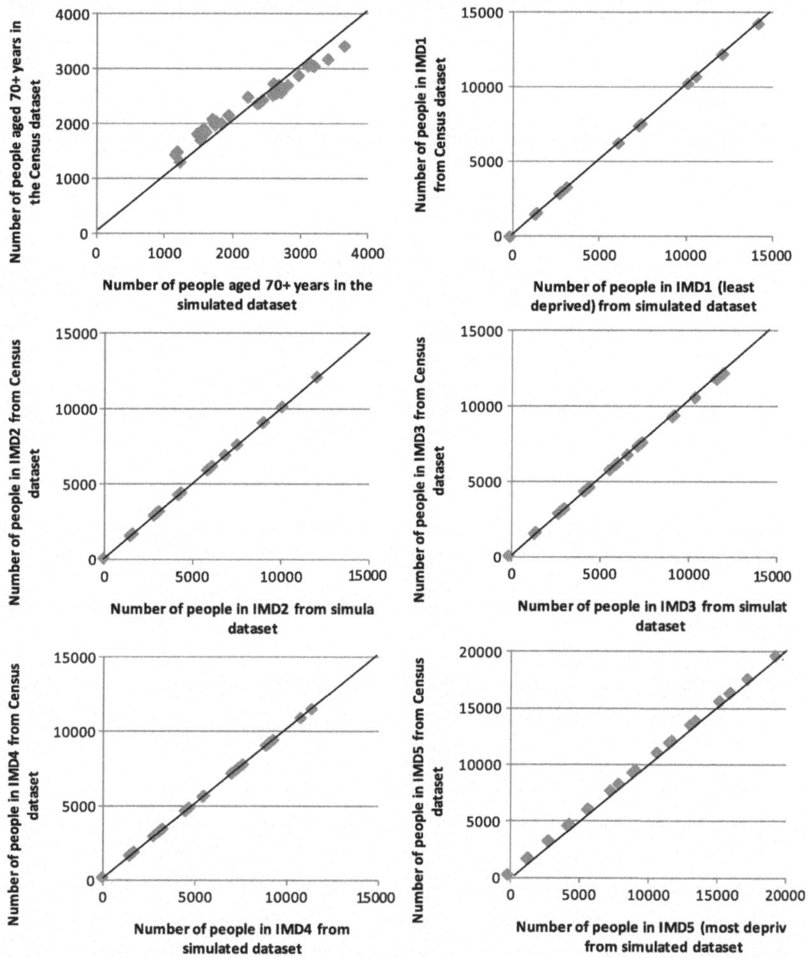

Fig. 5.3 (continued)

The simulated data were then used to examine the existence, and location, of any LSOAs with particularly high- or low-obesity prevalence, which would be useful information for deciding where to target a weight loss/healthy behaviour campaign. Alternatively, this may also be useful for deciding which areas to undergo further investigation in order to find out why those residents were particularly successful in managing their weight status and whether this translated into lower incidence of obesity related co-morbidities, such as diabetes or high blood pressure.

Accordingly, the prevalence of obesity (the number of obese residents in an area divided by the total adult population for that area) was mapped at LSOA level using ArcGIS (version 3) (see Fig. 5.4). The boundary data for these maps were downloaded from the UKBORDERS website (http://edina.ed.ac.uk/ukborders/). Figure 5.4 clearly shows that prevalence of obesity is not uniform across the study area, with higher rates in both urban and rural areas across the north of England.

Table 5.2 Results from the internal validation: R^2 from the linear regression models and *P* value from equal variance two-tailed *t* test

Variable	R^2	*P* value
Male	0.9941	0.7009
Female	0.9932	0.6965
0–15 years	0.9159	0.5591
16–19 years	0.9683	0.8090
20–49 years	0.9725	0.8093
50–69 years	0.9610	0.6200
70+ years	0.9739	0.4912
Deprivation quintile 1 (least deprived)	1.0000	0.9993
Deprivation quintile 2	1.0000	0.9989
Deprivation quintile 3	1.0000	0.9986
Deprivation quintile 4	1.0000	0.9985
Deprivation quintile 5 (most deprived)	1.0000	0.9988

Table 5.3 Comparison of the obesity prevalence for the simulated data versus the national figures for England (using data from the Health Survey for England for 2004, 2005 and 2006) by sex

		Data for England	Estimated data	Difference
All adults	% Obese	23.5	23.7	0.1
	% Overweight and obese	61.4	57.0	−4.4
Male	% Obese	23.0	21.5	−1.6
	% Overweight and obese	66.4	59.2	−7.1
Female	% Obese	24.0	25.6	1.6
	% Overweight and obese	56.4	55.1	−1.4

To explore these patterns further, a spatial scan statistic software, SaTScan (Kulldorf 2006), was used to determine whether any statistically significant areas of high prevalence ('hot spot') or low prevalence ('cold spot') existed across the study area. This model adjusts for population density to verify whether any clusters are over and above that expected based on population distribution. Two Bernoulli models were run examining the number of obese individuals, in one instance looking for any 'hot spots' of high prevalence and the other for 'cold spots' of low prevalence for each area (i.e. the 4,318 LSOAs). Both models required a 'case' file composed of the number of obese adults living in each area, a 'control' file being the balance of the population living in each area and a 'geo' file, which consisted of the Cartesian coordinates for the centroid of each micro-area. The model used the least subjective option for the maximum spatial cluster size, setting it at 50 % of the risk population. Further clusters needed to be distinct. No cluster boundaries were permitted to overlap geographically.

The output from these models is presented graphically in Figs. 5.5 and 5.6. Figure 5.5 shows the relative risk of obesity for each LSOA. A relative risk of 1.2 would mean that residents in that LSOA are 1.2 times more likely to be obese than the average for the study area. Similarly, residents with a relative risk of less than one are less likely to be obese. It can be seen that residents in some areas have an

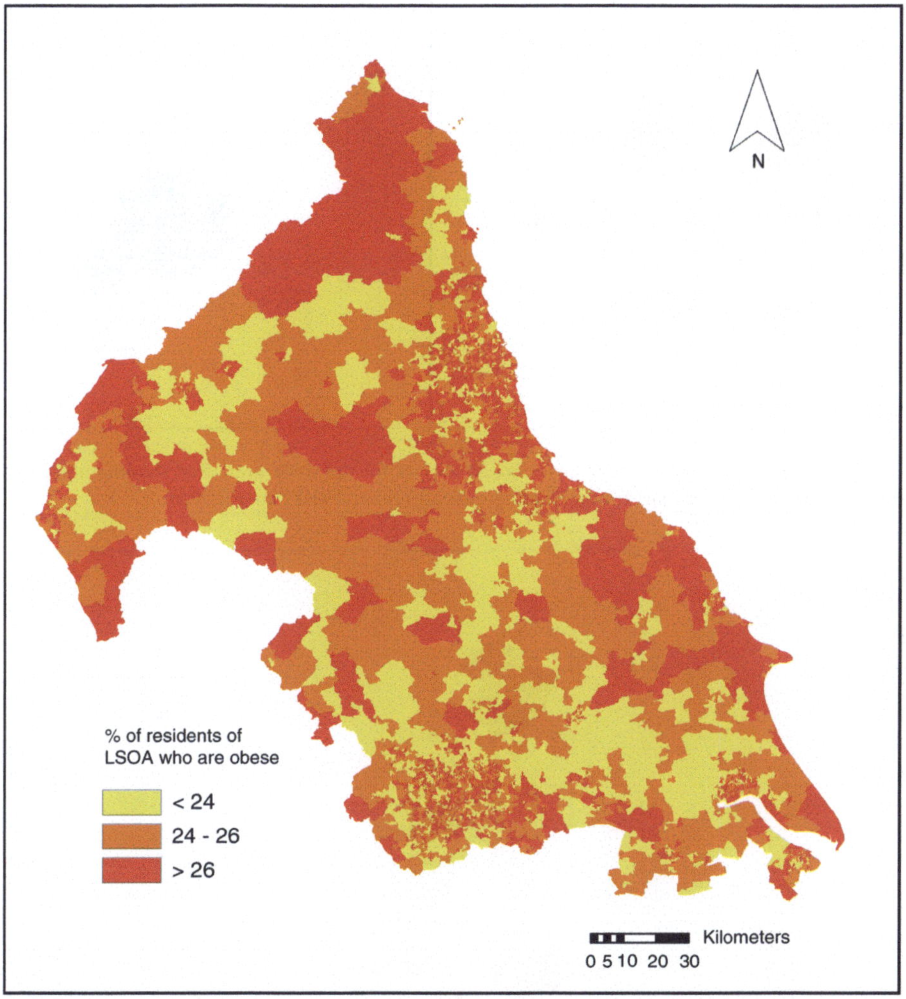

Fig. 5.4 Maps of adult obesity prevalence at LSOA level. Red shading indicates higher prevalence

increased risk of being obese compared to other areas. Note that this does not distinguish between contextual and compositional reasons for this clustering – thus whether obese people congregate within the same locality or whether there is an environmental impact of living in that neighbourhood. So what causes the residents to be/become obese is beyond the scope of this chapter.

Figure 5.6 shows the locations of the statistically significant clusters of either hot or cold spots of obesity. We clearly see how clusters of higher than expected (given the population density) prevalence of obesity are found in areas to the north-east, north-west and south of the study area. Conversely, cold spots are evident in the central section.

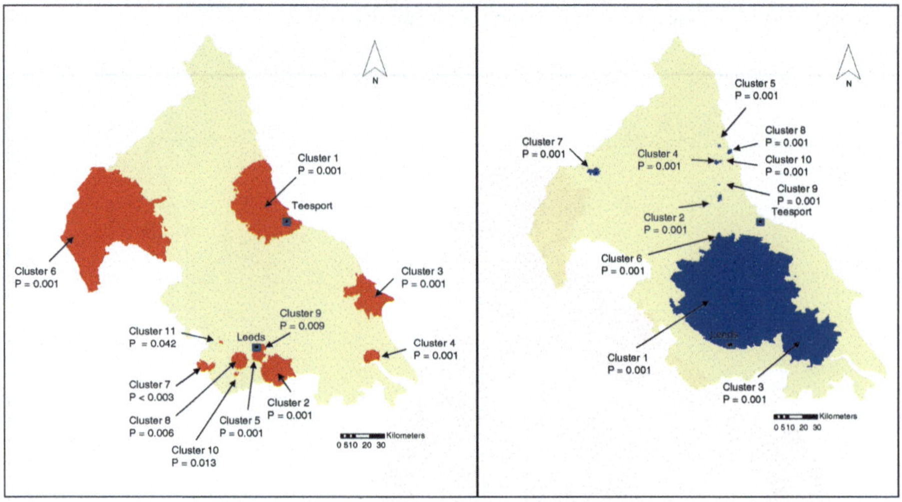

Fig. 5.5 Map of the clusters of relative risk of obesity for the study area. Red shading indicates relative risks are greater than 1; blue shading indicates relative risks are less than 1

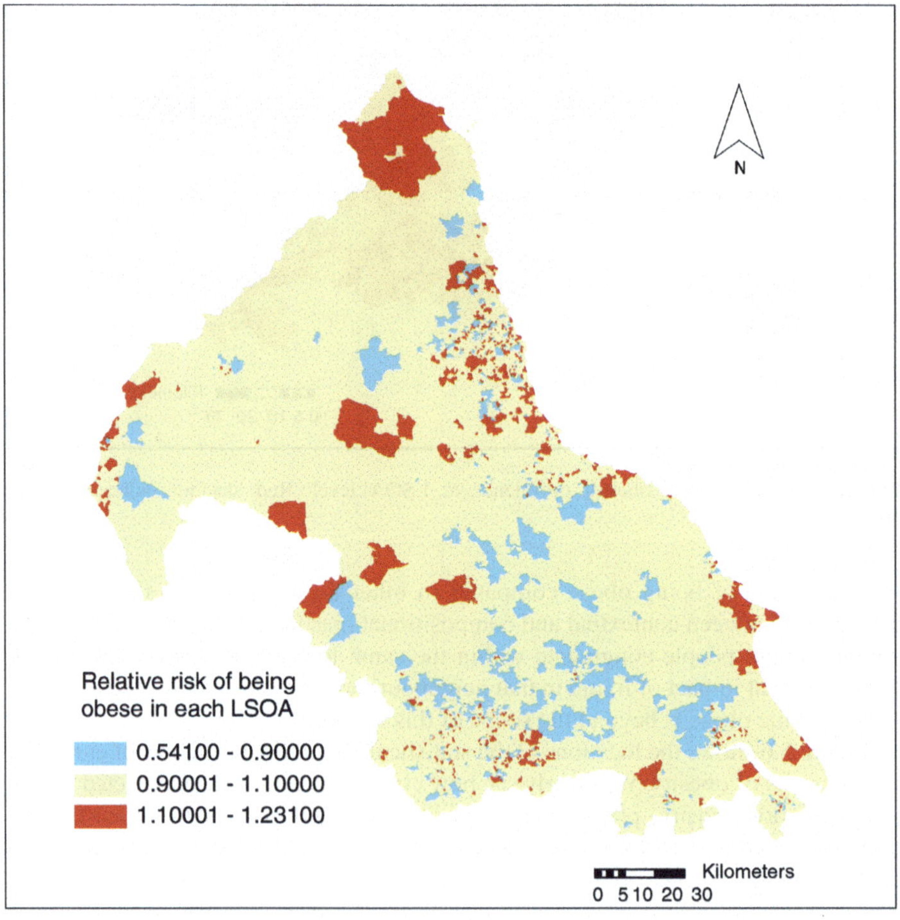

Fig. 5.6 Map of the areas of 'hot' and 'cold' spots of obesity prevalence at LSOA level

5.5 Conclusions

This chapter has outlined the algorithms behind SimObesity, a deterministic spatial microsimulation model which uses a combinatorial optimisation technique, with both a reweighting and integerisation algorithm. This model can be used to estimate new information at the micro-level and has been successfully used to estimate a variety of health data (Tomintz et al. 2008; Procter et al. 2008; Edwards et al. 2010b; Morrissey et al. 2010). To carry out a detailed survey to obtain the same extensive data would be extremely time consuming and expensive.

SimObesity can be used to construct synthetic small-area populations that are comprised of a wealth of socio-economic and health variables by linking data from different sources. This produces a new micro-level population dataset that is not available from published sources. These data can then be utilised to investigate associations between these variables at the local level.

Accordingly, we also show an application of the model to estimate adult obesity across the north of England. The validation of these data shows that the simulated data corresponds well with the actual data. The model is able to provide good estimates for all of the input variables, with no significant differences between the simulated and actual populations. The obesity estimates were comparable with the sample obesity data for the UK. The output data were then used to isolate statistically significant hot and cold spots of obesity prevalence. This information increases our understanding about obesity in the area, highlighting that prevalence is not uniformly high (or low) across the north of England (a simple deprivation-type distinction), and some areas require more investigation than others.

Spatial microsimulation is by its very nature both an art and a science. Modelling decisions may impact the resulting dataset. Having a large population dataset will facilitate a more accurate simulation, as it provides more individuals to select from the simulated population, thus making it more likely that some rare population characteristics are available for selection (science!).

However, the choice, and number, of input (constraint or benchmark) variables will also impact the final output (art!). For optimum simulation accuracy (i.e. for the attributes of the actual and simulated population to correspond), the input variables should be strongly correlated (ideally, $r>0.5$) with the output variables. Given there is not a linear relationship between predictive power and effect size, the predictive value of smaller correlation coefficients is not useful or reliable. Thus, this decision is best informed through a combination of relationships articulated in the literature and the strength of correlations between potential input variables and the proposed outcome variable(s) in the population dataset. While the original version of SimObesity was limited to a maximum of six input variables due to computational restrictions, programming improvements have removed this cap so any number of constraints may be utilised, either univariate or cross-tabulated or a combination of these. The only limitation therefore is that with too many input variables (or categories of the variables), the model could run into small number problems. Thus, if the Census dataset had zeros or low numbers for any input categories, the accuracy of the simulation is likely to be lower.

A further difficulty is in the validation of the simulated data. A simple comparison of the estimates to actual data is generally not possible as, by definition, these data do not already exist at the micro-level (otherwise, the simulation would not be necessary). Many chapters of this book suggest ways to estimate the error in the data to seek to achieve validation of these models, and Chap. 15 tries to summarise all these validation methods. Likewise, we advise the use of linear regression and paired unequal variance t tests to assess the precision and accuracy of the model estimates.

In conclusion, the reweighting deterministic algorithm in this chapter has been shown to be a useful tool for simulating micro-area data. However, spatial microsimulation models, although conceptually straightforward to design, are not simple to construct. Thus, this handbook highlights what models are available and which could potentially be used to fit/model your data.

References

Ballas, D., Rossiter, D., Thomas, B., Clarke, G. P., & Dorling, D. (2005). *Geography matters: Simulating the local impacts of national social policies*. York: Joseph Rowntree Foundation.

Birkin, M., & Clarke, G. P. (2011). The enhancement of spatial microsimulation models using geodemographics. *Annals of Regional Science*. doi:10.1007/s00168-011-0472-2.

Craig, R., & Mindell, J. (2008). *Health Survey for England 2006*. The NHS Information Centre. National Statistics. http://www.ic.nhs.uk/statistics-and-data-collections/health-and-lifestyles-related-surveys/health-survey-for-england/health-survey-for-england-2006-latest-trends. Accessed Aug 2011.

Communities and Local Government. (2010). English indices of deprivation: Consultation. ISBN: 978-1-4098-2413-8. http://communities.gov.uk/documents/localgoverment/pdf/1524728.pdf. Accessed Aug 2012.

Edwards, K. L., & Clarke, G. P. (2009). The design and validation of a spatial microsimulation model of obesogenic environments in Leeds: SimObesity. *Social Science and Medicine, 69*, 1127–1134.

Edwards, K. L., Clarke, G. P., Thomas, J., & Forman, D. (2010a). Internal and external validation of spatial microsimulation models: Small area estimates of adult obesity. *Applied Spatial Analyses and Policy*. doi:10.1007/s12061-010-9056-2.

Edwards, K. L., Clarke, G. P., Ransley, J. K., & Cade, J. E. (2010b). The neighbourhood matters: Studying exposures relevant to childhood obesity and the policy implications in Leeds, UK. *Journal of Epidemiology and Community Health, 64*(3), 194–201. doi:10.1136/jech.2009.088906.

Gatrell, A. C. (2002). *Geographies of health: An introduction*. Oxford: Blackwell.

Huang, Z., & Williamson, P. (2001). *A comparison of synthetic reconstruction and combinatorial optimisation approaches to the creation of small-area microdata* (Working Paper 2001/02). Department of Geography, University of Liverpool [online]. http://pcwww.liv.ac.uk/~william/microdata/workingpapers/hw_wp_2001_2.pdf. Accessed 13 Aug 2007.

Kulldorf, M. (2006). *SatScan user guide for version 7*. http://www.satscan.org/. Accessed Dec 2009.

Morrissey, K., Hynes, S., Clarke, G. P., & O'Donoghue, C. (2010). Examining the factors associated with depression at the small area level in Ireland using spatial microsimulation techniques. *Irish Geography, 45*(1), 1–22.

National Centre for Social Research and University College London. (2004). *Department of Epidemiology and Public Health, Health Survey for England, 2002* [computer file]. Colchester: UK Data Archive [distributor], 2004. SN: 4912.

Procter, K. L., Clarke, G. P., Ransley, J. K., & Cade, J. E. (2008). Micro-level analysis of childhood obesity, diet, physical activity, residential socio-economic and social capital variables: Where are the obesogenic environments in Leeds? *Area, 40*(3), 323–340.

Smith, D., Pearce, J. R., & Harland, K. (2011). Can a deterministic spatial microsimulation model provide reliable small-area estimates of health behaviours? An example of smoking prevalence in New Zealand. *Health and Place, 17*, 618–624.

Tomintz, M. N., Clarke, G. P., & Rigby, J. E. (2008). The geography of smoking in Leeds: Estimating individual smoking rates and the implications for the location of stop smoking services. *Area, 40*(3), 341–353.

Wilkinson, P., Grundy, C., Landon, M., & Stevenson, S. (1998). GIS in public health. In A. Gatrell & M. Löytönen (Eds.), *GIS and health*. London: Taylor & Francis.

Chapter 6
Spatial Microsimulation Using a Generalised Regression Model

Robert Tanton, Ann Harding, and Justine McNamara

This chapter outlines a method of spatial microsimulation that uses a reweighting algorithm implemented with the SAS programming language. The reweighting algorithm derives population weights by benchmarking the unit record level survey data to the reliable spatially disaggregated census data. These weights can then be applied to the sample to derive final populations for the small area, just like survey weights provided by national statistical agencies allow aggregation to national totals.

This chapter describes in detail the data used, the estimation methodology, and the advantages and disadvantages of the generalised regression method. An application to poverty estimation in Australia is also presented. Tanton et al. (2011) provide additional detail on the development of the model and applications of the model.

6.1 Data Sources

Any spatial microsimulation method that uses reweighting is going to require a survey dataset that contains the variables that need to be benchmarked, the variables for which small area estimates are required and a reliable small area dataset to reweight the survey dataset to.

The small area dataset must contain reliable small area data for a number of variables that are relevant to the variables for which small estimates are required. An example is using reliable data of the number of people in certain income groups, rent payment categories and mortgage payment categories for small areas to get

R. Tanton (✉) • A. Harding • J. McNamara
National Centre for Social and Economic Modelling, University of Canberra, Canberra, Australia
e-mail: Robert.tanton@natsem.canberra.edu.au

R. Tanton and K.L. Edwards (eds.), *Spatial Microsimulation: A Reference Guide for Users*, Understanding Population Trends and Processes 6, DOI 10.1007/978-94-007-4623-7_6,

estimates of housing stress (calculated using a measure of the ratio of housing costs to income), which are not normally available for small areas due to data limitations.

These reliable small area estimates are usually available from a census of the population or from administrative data, and the tables used are called benchmark tables, as they are the tables that the reweighting process used is trying to hit. These tables will all be data related to the number of people or households in the small area. An example of a benchmark is the number of people with incomes between $30,000 and $50,000.

The survey that is being reweighted must have variables matching the reliable small area data (see Chaps. 2 and 5). The survey dataset should also contain, or have enough information to calculate, the variable or variables for which small area estimates are required. As an example, if we want to generate small area estimates of poverty rates, the survey data will need to contain a variable capturing household disposable income, along with data about the number of people in the household so that equivalised income can be calculated. That is, this procedure adjusts the household disposable income for the number of adults and children in that household using the modified OECD scale.

For the spatial microsimulation model discussed here, we use survey data from the Australian Bureau of Statistics (ABS) and small area census data, also from the ABS. The model needs unit record data from the survey, and this is available through a confidentialised unit record file (CURF).

6.1.1 The ABS Survey of Income and Housing

The survey dataset used for this analysis was the Survey of Income and Housing (SIH), and we combined two survey years to get better estimates. The survey years used were 2002/2003 and 2003/2004. While there are later surveys available in Australia, these two were used in the version of the model (SpatialMSM08c) described in this chapter.

The Survey of Income and Housing is one of Australia's best sources of information on income and expenditure, essential when trying to model poverty rates and housing stress. Other possible sources of data include the Household, Income and Labour Dynamics Survey of Australia (HILDA); however, this is a longitudinal survey and therefore not as good as the SIH for the purpose of estimating poverty in 1 year. This is because the sample is smaller than the SIH, and the sample does not change over time.

The 2002/2003 SIH is a survey of 10,211 households with 19,402 persons aged 15 years and over (ABS 2004), and the 2003/2004 SIH is a survey of 11,361 households with 22,315 persons aged 15 and over (ABS 2005). For both files, a confidentialised unit record file (CURF) is available from the ABS on a CD-ROM.

Because the surveys were from different years, and were then being benchmarked to 2006 census data, we needed to inflate any financial data (incomes, housing costs, etc.) in each of the surveys to 2006 prices. We also needed to group some of the categories as the classifications between the surveys and the census were different. This process is outlined in Chap. 2.

The ABS provides the survey data in three files: a person-level file, a household-level file and an income unit file. An income unit is one person or a group of related people within a household who share incomes. This income sharing is assumed to take place between married couples and parents and dependent children, and income units can be seen as broadly equating to families. A household is a group of related or unrelated people living in the same dwelling.

To be able to benchmark the survey data to the 2006 census small area data, the data from all three levels are brought together into one file. This is done by matching each of the files by the Household ID, which is on every file. This matching creates a new person-level file with household and family data. For every person within the same household, the household characteristics are the same, and for every person in the same family, the family characteristics are the same.

We also needed to impute records for children onto the survey dataset. This is because the information for children under 15 is attached to the household record; thus, there is no person record for children under 15. In the census data, we did include children in some of our benchmark tables, so we needed to have something on the survey to be reweighted to these totals. We created a record for children from information on the household record and then imputed age and sex using the overall probabilities that a child is of a particular age and a particular sex. We then allocated the child other relevant household characteristics, such as household income.

We also needed to impute people in non-private dwellings onto the survey dataset, as these were not in the original scope of the survey but they are in the census dataset. We did this by using a 1% sample of 2001 census data. This process is described in Cassells et al. (2010).

The final file used for reweighting to the census data, after all these adjustments, has been called the 'SIH Linkage' file, as it is used to link the census to the survey data.

6.1.2 The 2006 Australian Census

The Australian Census of Population and Housing is conducted by the ABS every 5 years, and it covers the entire resident population of Australia. Cross-tabulations using the census data are available from the ABS, and these cross-tabulations can be derived for relatively small areas. The small areas used for this model were Statistical Local Areas (SLAs). This geographical unit was chosen from the ABS Australian Standard Geographical Classification (ASGC) because it was the smallest unit with complete coverage of Australia that did not introduce the problems of data confidentiality evident at smaller area levels such as census Collection Districts.

There were 1,426 SLAs in Australia in 2006, ranging in population from 12 to 181,327 people. These were distributed unevenly across Australia, with some small

Table 6.1 Example of a benchmark table

Small area	Income category	Rent category	Number of people
1	$30,000 – <$50,000	$200 – <$250	800
1	$50,000 – <$100,000	$200 – <$250	700
1	$30,000 – <$50,000	$250 – <$300	1,000
1	$50,000 – <$100,000	$250 – <$300	1,100
2	$30,000 – <$50,000	$200 – <$250	1,200
...			

states and territories being broken into relatively large numbers of SLAs and other larger states consisting of relatively few. For example, the Australian Capital Territory, which contains less than 2% of Australia's population, had 109 SLAs (or almost 8% of SLAs), while New South Wales, which contains 34% of the total population, had only 200 SLAs (or just over 14% of all SLAs). Of particular note was Queensland, which was divided into 479 SLAs, many of them in Brisbane and with quite low populations. Queensland thus contained 19% of Australia's population, but almost 34% of all SLAs.

All Australian census data is provided either for usual residents or as enumerated. The 'as enumerated' data is based on where a respondent was located on census night. Usual residence data places the respondent back in the area in which they usually reside for the purposes of an area-level population count. Because the survey data used for the model refers to where the respondent usually resides, the census data used is also usual residence data.

However, there are issues with the usual residence data that means it needs to be further adjusted to match the survey data. The usual residence data for people not at home on census night relates to the household information of where the person was enumerated on census night, not the area where they usually reside. So while the person can be (and is) moved back to their area of usual residence for the usual residence census count, the household characteristics are still for the house they were visiting on census night. This affects a fairly small proportion of the Australian population (about 4.7% of the population were not enumerated at home on census night), and these people were removed from the census benchmark tables, as their household characteristics (on which much of our benchmarking are based) do not reflect their true residential situation.

The other process required to get the census and survey data comparable was to take into account the 'not stated' values in the census. These are records where the respondent has not provided information on a particular variable. If this occurs in an ABS survey, the figures are imputed, but these values are not imputed on the census. For our purposes, we have used a pro rata method to allocate the not stated categories out to valid categories.

This census data is then used to create a number of benchmark tables, which are described below. These tables all consist of data related to the number of people in the small area in each category of the benchmark. An example of a benchmark table is shown in Table 6.1.

It can be seen that the benchmark tables are cross-tabulations, so the benchmarking process relates to marginal totals. The example shows only two categories of income and rent, but in reality there are many more, so the tables are quite complex and very large when all areas are combined.

Within each of these tables are a number of benchmark classes, which are the two income categories above ($30,000–<$50,000 and $50,000–<$100,000) and the two rent categories ($200–<$250 and $250–<$300).

This is an example of only one benchmark table. There are a number of these tables in the estimation process, and the weights derived for each household in the survey (the final 3 columns in Table 6.3) are designed to provide the best estimates for all the tables.

The model runs through each benchmark in turn. As more benchmarks are added to the procedure, the model struggles to match all of the benchmarks for all areas; thus, some areas are lost due to the procedure not converging. However, the areas that do still converge are more accurate because there have been more benchmarks utilised in determining the estimate.

This means that in selecting benchmarks, there is a trade-off between selecting more benchmarks and getting better estimates but having more areas for which it is not possible to produce a reasonable estimate, and using fewer benchmarks and getting more areas with reasonable estimates but having final estimates that are not as accurate. An ideal set of benchmarks is one that does not lose any areas that results are required for, but provides reasonably accurate results.

Benchmarks also need to be correlated with the final estimate being derived. In this chapter, we are deriving an estimate for poverty rates, so the benchmarks need to be correlated with poverty rates. One way to identify which variables to use may be to conduct a logistic regression of the likelihood of being in poverty and potential benchmark variables to identify which ones are most strongly associated with poverty. Alternatively, using relevant literature about the relationship between poverty and the types of benchmark variables available in the datasets is an alternative approach to identifying appropriate benchmark variables.

The benchmarks used for the model used as an example here were all chosen based on a review of the literature on poverty. They include housing tenure, as public housing tenants tend to have lower incomes (Queensland Council of Social Services 2009); family type, as different families experience different poverty rates (Buddelmeyer and Verick 2007; Miranti et al. 2011); and household income, which is used as the basis for calculating the poverty line. Other variables that are associated with poverty include labour force status, age, long-term disability, years of work and university education (Buddelmeyer and Verick 2007). Some of these variables cannot be included in the list of benchmarks because they are not available on both the survey and the census (a prerequisite for benchmark selection).

We included other benchmarks to ensure the process used for calculating poverty rates per person was based on accurate information. The benchmark table included for this was the number of people usually resident in the household. We also included a variable capturing the number of dwellings and a non-private dwelling benchmark, which is particularly relevant to estimating poverty as people in non-private

Table 6.2 Benchmark tables used in order used

Benchmark table	Description	Type	Number of benchmark classes
1	Age by sex by labour force status	Person	32
2	Number of occupied private dwellings	Household	1
3	Dwelling tenure by weekly household rent	Household	6
4	Dwelling tenure by household type	Household	15
5	Dwelling structure by household family composition	Household	24
6	Household size – number of persons usually resident	Household	6
7	Monthly household mortgage by weekly household income	Household	12
8	Different types of non-private dwelling	Person	4
9	Dwelling tenure by weekly household income	Household	25
10	Weekly household rental by weekly household income	Household	20
Total			130

Table 6.3 Example of a reweighted survey file

Unit record	Household ID	Weekly income	Weekly rent	Other variables	Household weight		SLA 1	SLA 2	Other SLAs
1	1	7	3	...	1,029		2.6	0	...
2	2	11	4	...	157		0	6.8	...
3	2	9	6	...	157		0	6.8	...
4	2	12	3	...	157	→	0	6.8	...
5	3	7	1	...	1,004		13.54	1.4	...
...									...
18,326	9,345				8,077,300		25,853	27,940	

dwellings (hospitals, nursing homes, and so on) often have different characteristics to people in occupied private dwellings and usually have higher poverty rates.

There were also other benchmarks included in order to help derive estimates of housing stress, which these weights were also used for. One of the advantages of this spatial microsimulation method is that the weights are generalisable, so they can be used for a number of output variables, depending on the benchmarks chosen. The weights derived using the benchmarks shown in Table 6.2 have been used to estimate both housing stress and poverty rates. The benchmark tables added to provide estimates of housing stress were measures of housing costs, with rent and mortgage payments separately benchmarked, and both dwelling structure (separate house, semi-detached, unit) and housing tenure (renting, owning, purchasing, and so on).

The benchmark tables were all cross-tabulations of the benchmark variables. These cross-tabulations mean that we are benchmarking to a number of variables together, which provides better estimates than just benchmarking to a single variable in a table. So looking at Table 6.2, we are benchmarking to the number of males aged 20–25 who are unemployed, rather than to all males, all people aged 20–25 and all unemployed people.

The list of benchmarks used in the model presented here is shown in Table 6.2. It can be seen that there are 10 benchmarks and a total of 130 benchmark classes. The most benchmark classes are in the first table, which is age by sex by labour force status with 32 classes.

The order in which benchmarks are entered into the model can affect the modelling results. In practice, the reweighting procedure places greater importance on the first benchmarks. The reason for this is complex and is described further in Tanton et al. (2011). In the model we present here, the benchmarks were used in the order they appear in Table 6.2.

6.2 Method

This model uses a deterministic reweighting methodology. It is based on the same method that the Australian Bureau of Statistics (ABS) used to reweight their surveys to national population totals, but in this case, we use the method to reweight survey data to small area population totals. The method is programmed in the SAS statistical programming language in a macro called GREGWT. The output from the procedure is the full survey dataset with a weight for each record and for each SLA. This weight can then be used to derive estimates for that SLA, in the same way that the survey weight would be used to derive national estimates.

A hypothetical example of the final survey dataset is shown in Table 6.3. The original survey weights are in the 'household weight' column, so summing these will give the total number of households in Australia. The weights for each SLA are shown as 'SLA 1', 'SLA 2', and so on for all SLAs. It can be seen that these weights are much smaller than the full survey weights, and there are more 0 weights for households that are not representative of any household in the SLA. The sum of all the SLA weights is the population for that SLA.

The reweighting method used is a generalised regression method. The method starts with an initial weight, which needs to be close to the final weight for that SLA, but is going to be further adjusted by the reweighting procedure. For our purposes, we use the household weight scaled by the population size as the starting weight. The general formula is

$$a_i = b_i * (\mathrm{Pop}_{\mathrm{sla}}/\mathrm{Pop}_{\mathrm{aus}})$$

where a_i is the new weight, i is the household on the survey, b_i is the starting weight (so the final weight from the original survey file for this household is scaled to the population of the SLA), $\mathrm{Pop}_{\mathrm{sla}}$ is the population for the SLA and $\mathrm{Pop}_{\mathrm{Aus}}$ is the Australian population.

In Table 6.3, for household 1, it would be

$$1{,}029 * (25{,}853 / 8{,}077{,}300) = 3.29.$$

In the basic linear regression method with no constraints, if x_i is a row vector of auxiliary variables for unit *i* on the survey, and *X* is a set of benchmark totals, then a first estimate of *X* using these initial weights (X_a) can be calculated as as $X_a = \sum_i a_i x_i$.

If X_c are the accurate benchmark totals from the census, then a new set of weights can be calculated from the first round of estimated benchmarks totals (X_a) as

$$w_i = a_i(1+(X_c - X_a)(\sum_i a_i x_i x_i')^{-1} x_i')$$

These weights meet the constraint $\sum_i w_i x_i = X_c$ while minimising the generalised least squared distance function $F^{GLS} = \sum_i (w_i - a_i)^2 / a_i$.

However, the procedure used for SpatialMSM uses a constrained distance function as there can be no negative populations, so all weights must be greater than zero. This constrained distance function is called the chi-squared distance function. This is minimised subject to the benchmark tables (X_c). The method is described in Singh and Mohl (1996), Bell (2000) and Tanton et al. (2011).

Because a boundary (the weight cannot be negative) is set, a truncated chi-squared distance function must be used which is

$$F^{chi} = \sum_i (w_i - a_i)^2 / a_i \quad \text{for } w_i \text{ in } [L_i, U_i].$$

This now requires an iterative approach as the boundary condition that $w_i > 0$ may not be met by the new weights.

For the first iteration ($m=0$), let $a_i^{(0)} = a_i$ for all *i* and $X_a^{(0)} = \sum_i a_i^{(0)} x_i$. An estimate for $A^{(0)}$ can then be calculated so that

$$X_c - X_a^{(0)} = A^{(0)} \sum_i a_i^{(0)} x_i x_i'$$

This can be simplified by letting $T^{(0)} = \sum_i a_i^{(0)} x_i x_i'$, so that the formula to be solved for $A^{(0)}$ is

$$X_c - X_a^{(0)} = A^{(0)} T^{(0)}$$

To simplify further, the matrix $T^{(0)}$ can be decomposed into U′U where U is an upper triangular matrix and U′ is the inverse of an upper triangular matrix. A solution for $A^{(0)}U'$ can then be calculated and then for $A^{(0)}$.

For each additional iteration ($m=m+1$), for each record *i*,

$$w^*_i = a_i(1+A^{(m-1)} x_i')$$

if $w^*_i < L_i$ then $w_i^{(m)} = L_i$, $a_i^{(m)} = 0$
Else if $w^*_i > U_i$ then $w_i^{(m)} = U_i$, $a_i^{(m)} = 0$
Else $w_i^{(m)} = w^*_i$, $a_i^{(m)} = a_i$
Similar to the first iteration (iteration 0),

$$X^{(m)} = \sum_i w_i^{(m)} x_i$$

A^* is calculated as a solution of

$$(X_c - X_a^{(m)}) = A^* \sum_i a_i^{(m)} x_i x_i'$$

This is calculated in the same way $A^{(0)}$ was calculated, involving a decomposition into a triangular matrix (U′U) and solving for $A^{*}U'$ and then A^{*}.

$A^{(m)}$ can then be calculated as

$$A^{(m)} = A^{(m-1)} + A^{*}$$

Convergence is achieved when either all boundary conditions are met or there is no improvement in the weights given the convergence criteria (ε) specified. For our purposes, the convergence criteria were set at 0.001. Thus, for each class in a particular benchmark (p), the iteration will stop when either the boundary conditions are met:

$$(|X_p - X_a^{(m)}{}_p|) < \varepsilon$$

or when there is no further improvement:

$$(|A^{(m)}_p - A^{(m-1)}_p|) < \varepsilon$$

or when the maximum number of iterations is reached. Otherwise, the process will repeat until one of these conditions is met.

In our model, we have set the maximum number of iterations to 30. We have experimented around this number, but anything above 30 gave little improvement in the estimates with considerable increases in running time. Communications with the author of the GREGWT procedure confirmed that there was little point in iterating too many times.

The indication of whether the procedure has converged in GREGWT depends on the number of iterations. If the maximum number of iterations is reached, the procedure stops and a non-convergence flag is set. In practice, we found that there were many SLAs that gave acceptable results with this convergence flag set – they were very close to converging when the maximum number of iterations was reached.

Because of this, we used a different criterion to identify the accuracy of the final estimate. This criterion looked at the sum of the absolute error for all benchmarks in the area and identified if this was greater than the population of the area. If it was, then the area was rejected as inaccurate. The formula is

$$\sum_p |(X_{cp} - X_{ap})| < \mathrm{Pop}^{sla}$$

where X_{cp} is the actual estimate for each benchmark class p from the census and X_{ap} is the modelled estimate for each benchmark class. This procedure is termed the accuracy criteria, as it is a measure of accuracy of the final weights in predicting the benchmark characteristics. In the model described in this chapter, there were no areas where the convergence criteria were met and the accuracy criteria were not met. There were 30 areas where the convergence criteria were not met but the accuracy criteria were met, so these areas were included, and there were 120 areas where both criteria failed, so these areas were excluded from the analysis. These 120 areas represented only 0.5% of the Australian population.

The very low proportion of the population that is lost through non-convergence of SLAs is due to the fact that the main reasons for the procedure failing are either when there are very few people living in an area or where the area has unusual characteristics, which make it very different from the survey data that we are using to represent the population. These areas tend to be remote indigenous communities in Australia, industrial areas where very few people live and areas with a high proportion of one type of population, like defence establishments and areas close to universities (areas which also often have relatively small resident populations). The problem with these areas is that the procedure has trouble benchmarking a survey representative of the Australian population to an area with a completely different type of population. The procedure works best for areas that have similar characteristics to the overall Australian population.

The final output from the procedure is a set of weights for every SLA that has passed the accuracy criteria. As these weights have been calculated by benchmarking the data to the 2006 census, they represent the population in 2006. If an estimate is required for a different year, then this can be done by inflating the weights using small area population estimates from another source. This method is described in Chap. 9.

The above procedure has been implemented in an SAS macro which is available from the Australian Bureau of Statistics, but could also be implemented in other programming languages. The procedure is run for every SLA and takes about 25 s per SLA. This means for an Australian run of 1,426 SLAs, the time taken is about 9 h. The final file of all records from the survey for all SLAs that have passed the accuracy criteria is about 281 MB, so the process is very resource and storage intensive.

Once the new regional weights have been calculated, they need to be applied back to the survey data that was used to calculate the weights. The first step in deriving some small area results is to calculate the variable of interest from the survey. This survey has to be the same survey used for the spatial microsimulation process, so that the small area weights available for each record on the survey can be matched to the variable of interest calculated for each record on the survey.

Note that for the model described in this chapter (SpatialMSM), two surveys were used, so the financial data in both surveys had to refer to the same period. This means that the financial data in both surveys had to be inflated to 2006, the year of the benchmark census data. The inflation factor used for incomes was the average weekly earnings inflator from the ABS.

The selection of benchmark tables was done with this final variable of interest in mind, and in this case, it was poverty rates. We have calculated poverty rates using a standard Australian approach which involves first removing any households with a negative or zero income from the dataset and then calculating a poverty line based on half the median disposable household income per person. A discussion of how poverty rates are calculated in Australia can be found in Tanton et al. (2009b) and Miranti et al. (2011).

For this chapter, we have used a fairly basic definition of poverty to show an application of this model. There is much discussion in the international literature on whether income measures of poverty are appropriate and how wealth and other variables contribute to poverty (see Alkire and Santos 2010; Tanton et al. 2010). There are also much more complex ways to calculate poverty which include the depth of poverty, and spatial microsimulation has been used to get these estimates of poverty for Australia (see Tanton 2011). However, this chapter's focus is the spatial microsimulation method, so we have used a simple income poverty measure.

The poverty rate for an area is the proportion of people in the area whose equivalised household income is below this poverty line (in this case half medium income). We have used the modified OECD method to equivalise income. Because we are interested in the number of people in poverty in the small area, the number of households can be multiplied by the number of people in the household, which can then be multiplied by the weight for that household to get the number of people in the SLA that the particular household in the survey represents. If the poverty flag is set for this household, then this weight is added to the total number of people in poverty in the SLA.

The formula for this is

$$\mathrm{Pov}^{\mathrm{sla}} = \sum_i A^{\mathrm{sla}}_i \, \mathrm{pov}_i \, \mathrm{Pers}_i$$

where i are the respondent households, A^{sla}_i are the final household-level weights for each household i for the SLA, pov_i is the poverty flag calculated for respondent household i in the survey (so it is a binary variable, 1 identifying a poor household and 0 otherwise), Pers_i is the number of people in household i and $\mathrm{Pov}^{\mathrm{sla}}$ is the number of people in poverty in the SLA.

The formula for the poverty rate is then

$$\mathrm{PovRt}^{\mathrm{sla}} = \mathrm{Pov}^{\mathrm{sla}}/\mathrm{Pop}^{\mathrm{sla}}$$

where $\mathrm{Pov}^{\mathrm{sla}}$ is the number of people in poverty in the SLA and $\mathrm{Pop}^{\mathrm{sla}}$ is the population in the SLA.

6.3 Results

Poverty rates for all SLAs in Australia are shown in Fig. 6.1. This map shows the quintile of the poverty rate into which each SLA falls. The quintiles are population weighted, meaning that there is an equal number of people in each quintile, not an equal number of SLAs in each quintile. The white areas on the map are those where the accuracy criteria failed, and it can be seen that many regions of Australia's Northern Territory have no estimates. While these areas have, as noted earlier, a very low population, many are likely to also be places with relatively high poverty rates, and their absence from our estimates is important to note when interpreting our results.

In Australia, the population of SLAs ranges from very small populations in remote areas to very large populations in urban areas. This means that more populous SLAs are likely to have a more heterogenous population, and less populous SLAs will have a more homogenous population. This is known as the modifiable areal unit problem, or the MAUP. This does mean that SLAs across Australia are not comparable. Normally, an adjustment is made to take into account this problem, and

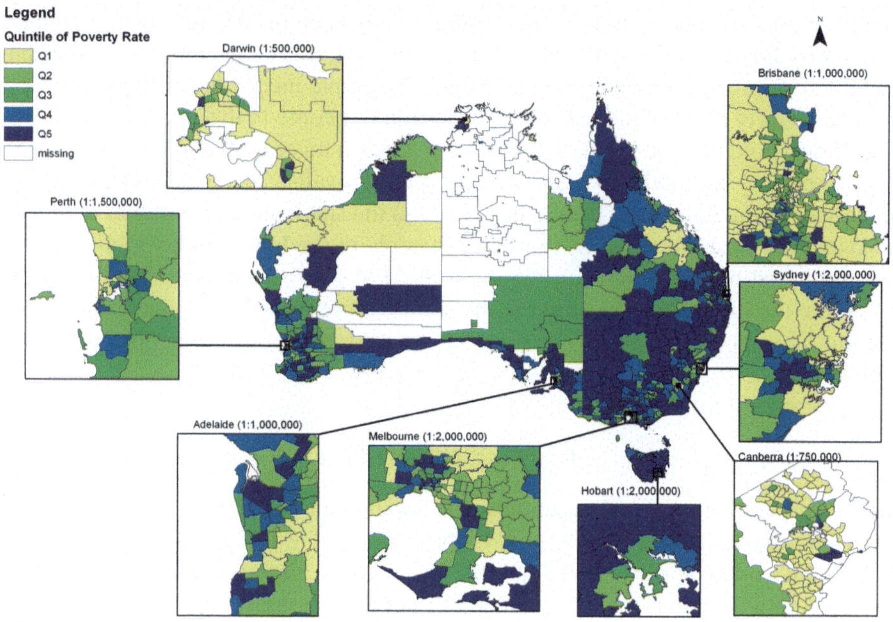

Fig. 6.1 Poverty rates (%) for all SLAs in Australia, 2006

this process is described further in Miranti et al. (2011). In this chapter, for simplicity, we have left all the areas at the SLA level.

It can be seen that poverty rates in Australia tend to be higher in remote and regional areas, with some areas of high poverty in the capital cities. There are pockets of low poverty in regional mining areas in Western Australia and north-east Queensland. Further analysis of Australian regional poverty rates is provided in Miranti et al. (2011) and Tanton et al. (2009b).

6.4 Validation

Validation is an important part of any small area model, and it tests how well the model replicates the real world. Chapter 15 outlines the different methods for validating models, and this section outlines the results from three of these methods:

1. Aggregate the model results to Australian state level and validate against poverty rates calculated from the 2005/2006 Survey of Income and Housing.
2. Apply the method outlined in 6.2 using state level benchmarks and compare to estimates from the 2005/2006 Survey of Income and Housing.
3. Compare proxy small area poverty rates that can be calculated from the census data to modelled estimates using the same definition.

The first method checks the model at a broad level (checking that we get the right totals for each state), but does not test the validity of the spatial distribution of poverty

Table 6.4 Validation of poverty rates using modelled poverty counts aggregated to state

	NSW	VIC	QLD	SA	WA	TAS	NT	ACT	Total
Survey of Income and Housing (SIH) 2005–2006 (%)	11.3	11.9	10.2	11.8	9.2	12.8	4.9	5.2	11
SIH 2005–2006 rank	4	2	5	3	6	1	8	7	
Spatial microsimulation (%)[a]	12.4+	11.7−	10.8+	13+	10.2+	14.6+	8.7+	6.1+	11.6+
Spatial microsimulation rank	3	4	5	2	6	1	7	8	

Source: ABS Survey of Income and Housing 2005–06; SpatialMSM/08B applied to 2002/03 and 2003/04 Survey of Income and Housing uprated to 2006

[a]+ (−) is when estimates from the spatial microsimulation are higher (lower) than the estimates for state level analysis directly from SIH 2005–2006

rates. The second method tests the estimation method for large areas, but does not test the spatial distribution of the small area results. The third method tests the spatial distribution of the results, but using a different definition of poverty. Overall, these three methods will provide a picture of how well the model works.

6.4.1 Aggregating Small Area Poverty Rates to State Level

Under this method, the number of people in poverty in each SLA is aggregated to state level. This aggregate figure is then converted to a percentage and compared to reliable state level estimates from the Survey of Income and Housing.

The results are shown in Table 6.4. It can be seen that poverty rates estimated from the spatial microsimulation model are generally higher than those calculated directly from the survey, but usually close. The slightly higher estimates could be due to the fact that the spatial microsimulation model is benchmarked to 2006 census data, which is then being compared to the 2005–2006 Survey of Income and Housing. These two data sources are collected in different ways (e.g. the census income is collected only in ranges, and all data is based on a self-completion form, rather than an interview) and for slightly different time periods, so it would be expected that slightly different numbers would come out of each dataset. When looking at the ranks, rather than the proportion of people in poverty, these are very similar.

The main difference in the ranks appears to be in Victoria, and the main reason seemed to be that where the model normally overestimates poverty rates, in Victoria, it underestimated them.

6.4.2 Reweight State Level Data

Using this method, the model was run at the state level, so the benchmarks (instead of being set at an SLA level) were all instead created as state level benchmarks. State level estimates of poverty calculated for 2006 were then compared to state level estimates of poverty from the 2005–2006 Survey of Income and Housing. If reliable

Table 6.5 Validation of poverty rates using modelled poverty counts calculated by state

	NSW	VIC	QLD	SA	WA	TAS	NT	ACT	Total
Survey of Income and Housing (SIH) 2005–2006 (%)	11.3	11.9	10.2	11.8	9.2	12.8	4.9	5.2	11
SIH 2005–2006 rank	4	2	5	3	6	1	8	7	
Spatial microsimulation (%)[a]	12.6+	11.7–	11.0+	13.2+	10.4+	14.6+	11.4+	6.4+	11.8+
Spatial microsimulation rank	3	4	6	2	7	1	5	8	

Sources: ABS Survey of Income and Housing 2005–06; SpatialMSM/08B applied to 2002/03 and 2003/04 Survey of Income and Housing uprated to 2006

[a]+ (–) is when estimates from the spatial microsimulation are higher (lower) than the estimates for state level analysis directly from SIH 2005–2006

results are achieved for each state, which can be readily checked against state level data from the survey, then this provides further support for the robustness of the model, with a higher likelihood that any errors for small areas will be due to the nature of the small areas, rather than any underlying errors in the modelling approach.

The reweighting was done at the state level for all benchmarks shown in Table 6.2, and the results are shown in Table 6.5.

It can be seen that the model estimates are still higher than the survey estimates, so the model based on the census data does provide higher estimates of poverty, even when results are produced based on larger geographical units. Looking at the ranks, the main differences seem to be in Victoria, which went from being ranked 2nd from the survey to being ranked 4th from the model, and the Northern Territory, which went from being ranked 8th in the survey to being ranked 5th using the model. Comparing these results to Table 6.4, it can be seen that the Northern Territory had a much higher poverty rate when the benchmark variables were at a state level. This could be because many of the smaller and high poverty areas in remote NT failed the accuracy criteria, so would be excluded from Table 6.4, but they will be included in Table 6.5 which is all calculated at the state level.

The other reason for this difference could be the ABS figures. The ABS sample for the SIH excludes remote areas in the Northern Territory, so poverty rates in the NT from the SIH could be low because very low-income people in remote areas are excluded from the ABS sample.

6.4.3 Compare to Small Area Proxy Poverty Rates from the Census

This method compares modelled small area poverty rates to census small area poverty rates using a different definition of poverty than is usually used, and one that can be estimated from the census.

The method for this technique is described in the validation chapter (Chap. 15). The proxy poverty rate which can be calculated from the census is gross income below $500 per week. A graph of the modelled proxy poverty rates and the proxy

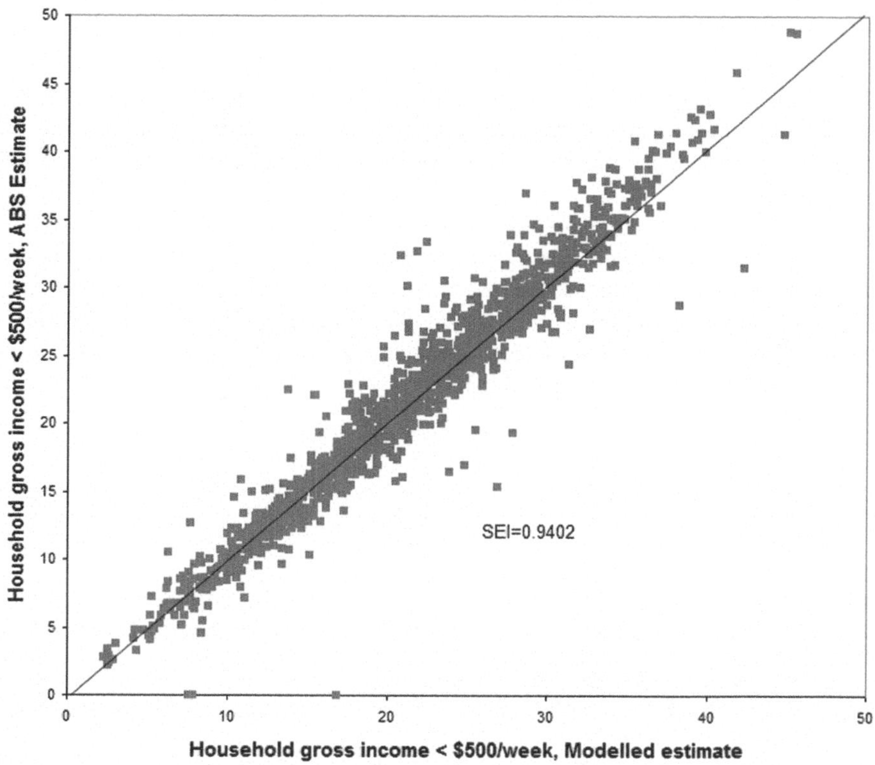

Fig. 6.2 Validation of proxy poverty rates by statistical local area, 2006

poverty rates estimated from census data is shown in Fig. 6.2. It can be seen that the standard error around identity (SEI) of 0.94 is very high and comparable to other SEIs calculated using this model (Miranti et al. 2011). One interesting point is that if a regression line was plotted, rather than the SEI, it would be very close to the 45 ° line shown on this chart. This suggests that the errors from the model are random, showing no strong bias in either direction.

Overall, the small area estimates from the model have been validated in three ways. All these methods of validation showed reasonable results, although the best results came from validating the data to small area 2006 census data. Aggregating the data in a number of ways showed some differences and biases, which may be because the validation was against a different survey, which was external to the spatial microsimulation model and would be expected to have different results.

6.5 Conclusions

This chapter has shown an example of a spatial microsimulation model being used in Australia for estimating poverty rates and housing stress. Validation of the method against comparable data using several techniques shows that it provides excellent results.

One of the real strengths of this approach to small area estimation is that the output is a reweighted survey file for two Australian surveys. Using these reweighted survey files with a tax/transfer microsimulation model based on the same surveys, policy effects can be modelled, and the small area impact of these modelled policy changes can be mapped. This provides a powerful tool for policy makers. Applications of this methodology have been published elsewhere (Harding et al. 2009; Tanton et al. 2009a).

References

ABS. (2004). *Household income and income distribution, 2002–03* (Cat # 6523.0). Canberra: ABS.

ABS. (2005). *Household income and income distribution, 2003–04* (Cat # 6523.0). Canberra: ABS.

Alkire, S., & Santos, M. E. (2010). *Acute multidimensional poverty: A new index for developing countries* (OPHI Working Paper 38). Oxford: OPHI.

Bell, P. (2000). *GREGWT and TABLE macros – Users guide*. Canberra: Australian Government (Unpublished).

Buddelmeyer, H., & Verick, S. (2007). *Understanding the drivers of poverty dynamics in Australian households: Institute for the study of Labor* (IZA Discussion Paper Number 2827). http://ftp.iza.org/dp2827.pdf. Accessed 18 Sept 2011.

Cassells, R., Harding, A., Miranti, R., Tanton, R., & McNamara, J. (2010). *Spatial microsimulation: Preparation of sample survey and census data for SpatialMSM/08 and SpatialMSM/09* (NATSEM Technical Paper 36). https://guard.canberra.edu.au/natsem/index.php?mode=download&file_id=1065. Accessed 18 Sept 2011.

Harding, A., Vu, Q. N., Tanton, R., & Vidyattama, Y. (2009). Improving work incentives and incomes for parents: The national and geographic impact of liberalising the family tax benefit income test. *Economic Record, 85*(s1), S48–S58.

Miranti, R., McNamara, J., Tanton, R., & Harding, A. (2011). Poverty at the local level: National and small area poverty estimates by family type for Australia in 2006. *Applied Spatial Analysis and Policy, 4*(3), 145–171.

Queensland Council of Social Services. (2009). *Poverty in Queensland 2009*. Brisbane: Queensland Council of Social Services. http://www.qcoss.org.au/sites/default/files/QCOSS%20Anti-Poverty%20Week%20Report%202010.pdf. Accessed 18 Sept 2011.

Singh, A. C., & Mohl, C. A. (1996). Understanding calibration estimators in survey sampling. *Survey Methodology, 22*(2), 107–115.

Tanton, R. (2011). Spatial microsimulation as a method for estimating different poverty rates in Australia. *Population, Space and Place, 17*(3), 222–235.

Tanton, R., Vidyattama, Y., McNamara, J., Vu, Q. N., & Harding, A. (2009a). Old, single and poor: Using microsimulation and microdata to analyse poverty and the impact of policy change among older Australians. *Economic Papers: A Journal of Applied Economics and Policy, 28*(2), 102–120.

Tanton, R., McNamara, J., Harding, A., & Morrison, T. (2009b). Small area poverty estimates for Australia's Eastern Seaboard in 2006. In A. Harding & A. Zaidi (Eds.), *New frontiers in microsimulation modelling* (pp. 79–97). Vienna: Ashgate.

Tanton, R., Harding, A., Daly, A., McNamara, J., & Yap, M. (2010). Australian children at risk of social exclusion: A spatial index for gauging relative disadvantage. *Population, Space and Place, 16*(2), 135–150.

Tanton, R., Vidyattama, Y., Nepal, B., & McNamara, J. (2011). Small area estimation using a reweighting algorithm. *Journal of the Royal Statistics Society Series A*. doi:10.1111/j.1467-985X.2011.00690.x.

Chapter 7
Creating a Spatial Microsimulation Model of the Irish Local Economy

Niall Farrell, Karyn Morrissey, and Cathal O'Donoghue

7.1 Introduction

There has been a growing emphasis on the spatial targeting of policy options in the area of poverty and social exclusion in Ireland since the early 1990s. As O'Donoghue et al. (2012) illustrate, the National Anti-Poverty Strategy (1997) has a spatial dimension in two of its five priority themes: disadvantaged urban areas and marginalised rural communities. Along with this, the recently updated National Spatial Strategy (2010) presents a national programme of development actions to reduce interregional inequality. Within these frameworks, local partnerships have been utilised as a mechanism to target resources at poverty "black spots" (Haase and Foley 2009). The importance of spatial policy such as this has been emphasised by findings that poor households tend to group together in specific areas (Jencks and Mayer 1990; Hajnal 1995; Ravallion and Jalan 1997). In light of this, policymakers would like to be able to identify the spatial context of poverty and/or target resources towards individuals/areas that need them the most (Watson et al. 2005; Tanton et al. 2009a, b; Vidyattama et al. 2011). The benefit of such a regional approach to welfare policy has been illustrated by Elbers et al. (2007).

Identifying the spatial incidence of welfare has been limited in an Irish context by the lack of effective spatially referenced income microdata, with studies to date confined to aggregate spatial disaggregations at county or regional authority level

N. Farrell (✉)
SEMRU, J.E. Cairnes School of Business and Economics,
NUI Galway, Galway, Ireland
e-mail: n.farrell1@nuigalway.ie

K. Morrissey
School of Environmental Sciences, University of Liverpool, Liverpool, UK

C. O'Donoghue
Rural Economy and Development Programme,
Teagasc, Athenry, Co., Galway, Ireland

R. Tanton and K.L. Edwards (eds.), *Spatial Microsimulation: A Reference Guide for Users*, Understanding Population Trends and Processes 6, DOI 10.1007/978-94-007-4623-7_7,

(e.g. O'Leary 2003; Morgenroth 2010). A number of aspatial microdata sources exist. Census microdata, known as the Sample of Anonymised Records (SARS), are available, but these data are unsuitable due to a lack of information on household composition and income whilst also employing an aggregate spatial scale. National accounts data present the most accurate representation of income, but these data are only available at the aggregate county level. The Living in Ireland (LII) survey contains income and employment information at the individual and household level. The 2000 dataset contained data on 13,067 individuals and information on a variety of individual, demographic and socio-economic characteristics, including income, employment and household composition statistics. However, LII data are only available at a coarse spatial scale. In considering spatially referenced data, Haase and Foley (2009) have noted that the only data detailing socio-economic population distributions at the local level are the small area census data. Known as the small area population statistics (SAPS), these data contain census information disaggregated to the electoral division (ED) level. The 3,440 EDs represent the most disaggregated spatial scale in Ireland. The population in any one ED ranges from a low of 55 individuals to a high of 14,238, with an average across all EDs of 885 (Morrissey et al. 2008). However, as with most censuses, data on income and welfare are limited. Merging LII data with the ED-level census data would create a spatially referenced micro-dataset containing estimates of Irish income, labour and welfare distributions at the local level. This would provide a much richer dataset at a very local level of spatial resolution.

In the absence of pre-existing spatial data, a spatial microsimulation model known as the Simulated Model of the Irish Local Economy (SMILE) was developed to synthesise regional distributions of welfare. This was carried out at the Teagasc Rural Economy and Development Programme (REDP), in collaboration with the School of Geography, University of Leeds.

It is the purpose of this chapter to give an insight into the rationale, development and application of SMILE in analysing the spatial incidence of welfare and income distribution in Ireland. This chapter continues as follows: Sect. 7.2 introduces SMILE, discussing the objectives of the simulation methodology relative to existing synthesis methodologies, whereby the requirement for a new synthesis procedure is motivated. This procedure, which we call quota sampling, is described in Sect. 7.3. The validation procedures employed to ensure an accurate synthesis are outlined in Sect. 7.4. In order to ensure that income distributions are aligned to known county-level distributions, a calibration procedure is employed. This is outlined in Sect. 7.5. Section 7.6 illustrates the application of SMILE to measure the spatial incidence of income redistribution in Ireland, and Sect. 7.7 provides some conclusions.

7.2 SMILE

SMILE is a static spatial microsimulation model, designed to simulate regional welfare, income and labour distributions and thus provide a basis for regional economic analysis in Ireland. As with similar international microsimulation models (e.g. Ballas et al. 2005a; Chin et al. 2005; Edwards and Clarke 2009), SMILE may be used to provide government, policymakers and non-government organisations with detailed spatial data which could be used to improve policymaking, analyse sectoral and regional investments and target resources.

A number of techniques exist which may be used by SMILE to synthesise data. Ballas et al. (2005b) provide an overview, with iterative proportional fitting (IPF) and various combinatorial optimisation (CO) methodologies being of greatest prominence. When deciding on which procedure to employ for SMILE, the primary objectives are the capacity to handle a combination of individual and household constraints and adequate run-time efficiency. The merits of existing procedures will now be discussed relative to these objectives.

Iterative proportional fitting (IPF – see Chap. 4) is a probabilistic methodology for constructing spatially disaggregated tables from aggregate spatial totals in the absence of pre-existing microdata. This is carried out by adjusting a two-dimensional matrix iteratively until row sums and column sums equal some predefined aggregate values, and in a geographical context, it can be used to generate disaggregated spatial data from spatially aggregated data (Wong 1992). For example, IPF may be used to create a disaggregated age/sex tabulation from separate age and sex small area totals (O'Donoghue et al. 2012). It has been found that IPF can potentially produce unrealistic data as probabilities are used to create synthetic microdata from regional aggregates, rather than using real survey data (Norman 1999). IPF was employed in the first version of SMILE, but a preference for using actual survey data motivated the adoption of a new approach. It was also found that difficulties arose when the unit of analysis of the constraint was different to that of the microdata. This is because IPF is designed to reweight individual-level microdata according to individual-level constraints. However, welfare analyses require that we synthesise at the household level, using microdata of individuals grouped into households. The IPF procedure finds it difficult to handle the additional degree of dimensionality imposed by reweighting individuals grouped into households according to individual-level constraints, and thus is unsuitable for SMILE.

Combinational optimisation (CO) techniques overcome the synthesis issues of IPF by reweighting existing survey microdata at either the individual or household level. CO techniques may be either deterministic (see Chap. 5) or probabilistic (see Chap. 3) in nature.

Deterministic reweighting assigns weights to each household based on the probability of that household belonging to the region in question (Ballas et al. 2005b). Similar to IPF, deterministic reweighting algorithms are computationally efficient. Such algorithms are unsuitable for SMILE, however, because they have multiple units of analysis (i.e. individuals grouped into households) which then require the use of non-trivial methods of weight generation, such as generalised regression weight-based methods. An example of such a generalised regression weight-based method is GREGWT, developed by the Australian Bureau of Statistics (Bell 2000) and used in the Australian spatial microsimulation model described in Chap. 6.

GREGWT uses a constrained distance minimisation function which uses a generalised regression technique to get an initial weight and iterates the regression until an optimal set of household or individual weights for each small area is derived

(O'Donoghue et al. 2012). Williamson (2009) highlights that when there are large numbers of constraints, the GREGWT algorithm does not always converge. Furthermore, such convergence issues are especially evident in areas of relatively low population density (Tanton et al. 2007). As many Irish EDs are of low population density (25% contain less than 100 households; 57% contain less than 200), significant barriers to convergence may exist if a generalised regression weight-based method such as GREGWT were to be used for SMILE.

The alternative to deterministic reweighting is probabilistic reweighting processes, the most popular of which is simulated annealing (SA). SA allows the survey data and constraints to have different units of analysis. Unlike IPF, SA contains mechanisms to avoid becoming trapped at local minima (Wu and Wang 1998). It is also less sensitive to convergence issues. Williamson (2009) found that in an Australian simulation, SA performed slightly better at matching than GREGWT for both constrained and unconstrained variables. This was particularly the case in districts where there was no convergence.

The main disadvantage of SA is the high computational intensity, which is due to the degree to which new household combinations are tested for an improvement in fit during simulation. To illustrate, Hynes et al. (2009) found that it took two days to generate almost 140,000 individual farm records from 1,200 survey data points on a 2 G Dell workstation. Scaling this computational requirement to a population of over four million people using a greater number of constraints, the simulation of SMILE may take a number of months. This is even more burdensome as it is desired to carry out repeated simulations for sensitivity analyses and simulations of future population projections.

Thus, practical restrictions imposed by great computational intensity have limited the application of SMILE under SA, motivating the development of a more efficient algorithm through a reduction in the number of required computations. We call this new process "quota sampling".

7.3 Quota Sampling

Quota sampling (QS) is a probabilistic reweighting methodology developed by Farrell et al. (2012). This procedure operates in a similar fashion to SA, whereby survey data are reweighted according to key constraining totals for each small area, with amendments made in the sampling procedure in order to improve computational efficiency. The basic sampling procedure, and its implementation in the overall simulation process, is outlined below.

7.3.1 Conceptual Overview

The quota sampling procedure analyses individuals grouped into households against constraints at either the individual or household level (see Fig. 7.1). Similar to SA, quota sampling selects observations at random and considers whether they are suitable

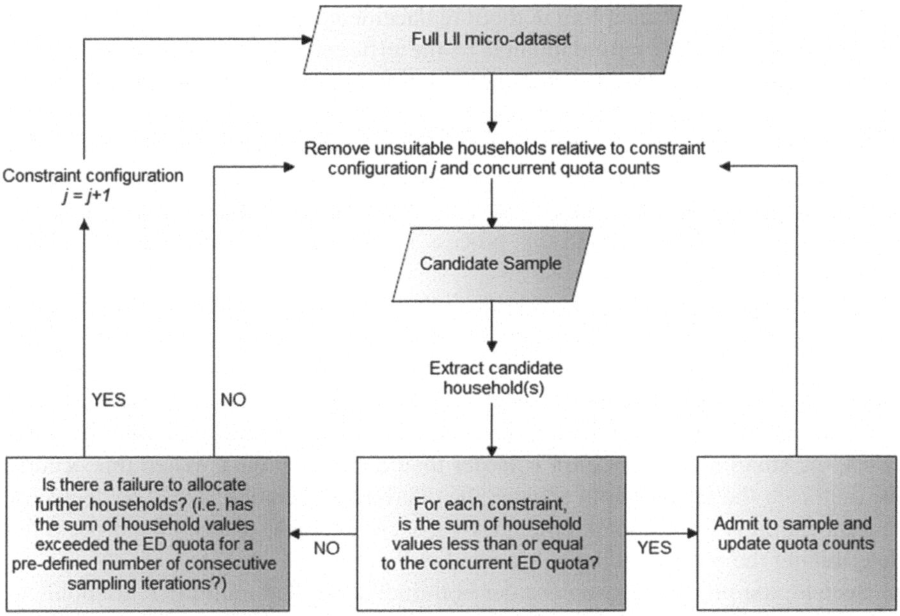

Fig. 7.1 Quota sampling synthesis procedure

for admittance to a given small area population based on conformance with aggregate totals for each small area characteristic. Unlike SA, quota sampling only assigns households that conform to aggregate constraint totals, and once a household is deemed selected, it is not replaced. To accommodate this, small area aggregate totals for each constraint variable are designated as the initial values for what we term "quotas". These quotas may be considered as running totals for each constraint variable, which are recalculated once a household is admitted to a small area population. The basic procedure is best explained in the context of allocating one household at a time, in the presence of a single age constraint. If the household sum of each constraining characteristic (e.g. two persons aged 20–25) is less than or equal to the small area total (e.g. ten persons aged 20–25), the household is assigned to the small area population. Upon deeming a household appropriate for a given small area, the quota counts are reduced by the sum of the characteristics of the assigned household(s). For individual-level constraints, we increment the running totals per constraint by the number of people in the household with that particular constraint. For household-level constraints, we increment by 1 (in our example, the ED quota would be amended to eight persons aged 20–25). This procedure continues until the total number of simulated individuals is equal to the small area population aggregates (i.e. all quotas have been filled). Thus, one can see that the intra-household variation of admitted households cumulates in a random sort which is consistent with aggregate constraint totals.

This mechanism of sampling without replacement avoids the repeated sampling procedure of SA and is fundamental to the efficiency gains of the quota sampling procedure. One can see that the process is analogous to the type of quota sampling undertaken by market researchers, whereby only individuals considered relevant to concurrent quota counts are admitted to a sample. This method of improving efficiency does present a number of convergence issues, however. A process analogous to the "swapping" of simulated annealing (Morrissey et al. 2008; Hynes et al. 2009) is discussed in Sect. 7.3.4, and this process is undertaken when constraint quotas approach capacity.

Quota sampling allows for further efficiency gains to be achieved at the implementation stage. First, households which do not comply with concurrent quota counts are extracted from the microdata population before each iteration of the sampling procedure. This limits the number of households to be considered to those relevant, reducing the size of the candidate dataset and improving computation time. In some cases, the small area population is larger than the survey data. When this occurs, we duplicate the microdata to achieve the district's population size. The degree of multiplication is subject to an efficiency vs. accuracy trade-off, and the process for determining the optimal point is discussed in Farrell et al. (2012).

Second, the procedure can consider both individual and multiple households in one simulation iteration. As stated, the candidate sample at each stage is limited to households eligible according to the quota counts at the initiation of the procedure. If we assign a number of households so that the total population assigned is less than or equal to the smallest constraining quota, we are assigning the maximum number of households in one iteration such that quota counts may not be exceeded, regardless of the distribution of characteristics. For example, if the smallest quota for a given ED is for 20 females aged 20–24, we can randomly assign multiple households in one iteration without exceeding any quota if the total population of the assigned group does not exceed 20 persons.

7.3.2 Implementation of QS in the Synthesis of SMILE

Having discussed the concept of QS, its implementation in SMILE is now described. As with all spatial microsimulation models, the initial consideration is that of choosing which variables constrain the data fusion (Smith et al. 2009). O'Donoghue et al. (2012) outline the process of choosing constraints in SMILE using bivariate regressions of candidate variables against disposable income in the LII microdata. In doing so, age, sex, level of education and household size are chosen. The additional constraint of household size is utilised in order to ensure an accurate distribution of household numbers per district. Once constraints have been decided upon, a number of practical limitations which can potentially prohibit convergence must be overcome during implementation. These issues and their corrective measures will now be outlined.

7.3.3 Practical Issues Prohibiting Convergence

Problems may arise in relation to the distributions of household size. The absence of an explicit constraint on the number of households allows for sampling without replacement to provide an accurate allocation of individuals. Smaller households contain fewer individuals and thus a smaller sum for the constraining criteria, making them easier to assign than larger households. This may result in a disproportionate amount of small households to be assigned per local area. This problem also affects the synthesis of households containing children. The nature of household structures requires children to be assigned alongside at least one adult. As smaller households which may not contain children are easier to assign, quotas for adults may fill before those for children. As no further children can then be assigned, this leads to a consistent under-representation of households containing children.

Furthermore, disparities in population distributions between census and survey totals may create a number of problems for household-based microsimulation procedures. This is because survey microdata are representative at the national level, whereas SAPS data are representative at the ED level. This poses little difficulty in simulating small areas that have a population distribution similar to that of the national distribution, but regions that differ from the national distribution may lead to some demographic groups consistently being under-represented in a given ED. Such deviations may be further increased if an ED contains individuals who live in institutions such as nursing homes, religious orders, psychiatric units, etc. (i.e. non-household members) as survey data generally do not cover individuals that are not part of a household. In the case of institutions such as boarding schools, children's hospitals or young offender's institutions, for example, we may have a situation where there are many children in an area relative to the number of adults. These differences may cause some EDs to consistently fail in reaching adequate convergence.

Finally, the use of sampling without replacement in quota sampling results in quota counts becoming increasingly more restrictive as the simulation progresses. As quota counts reach their target, the search space is continuously refined in accordance with concurrent quotas, whereby all households no longer eligible given updated quota totals are removed from the subset and the procedure is repeated.[1] When each constraint allocation reaches its target quota, all individuals with that characteristic are removed from the candidate search space. These mechanisms cumulate to offer a continuously diminishing search space and may prohibit convergence, whereby no household is able to satisfy all concurrent quota counts.

---

[1] For example, with a remaining quota count of n individuals of class k to be filled, the search space is refined to exclude households containing $n + 1$ individuals of class k.

Table 7.1 Ordered simulation procedure

Configuration stage	Constraint configuration *j*	Description
Correcting for under-representation	1	Random sampling of households containing children/random sampling of large households
Further correcting for under-representation	2	Random sampling of under-represented households
Basic sampling procedure	3	Random sampling of all households
Broadening of constraints	4	Removal of constraints, one at a time

7.3.4 Corrective Measures

The problems identified in Sect. 7.3.3 are corrected by an ordered simulation procedure whereby *j* constraint configurations are specified. Each j constraint configuration addresses an issue prohibiting convergence, determined by three criteria: the number of constraints consistently under-represented across all EDs, the number of constraints disproportionate to national disaggregations and the number of broadening criteria. These findings result in the creation of the ordered simulation procedure outlined in Table 7.1.

This order is determined by the requirements for accurate simulation. Taking configuration 3 as the basic sampling procedure, the primary decision to be made is whether to introduce the other steps before or after this stage. Configurations 1–2 are carried out beforehand as it is required that under-represented households be assigned before all others to correct for consistent under-representation. This consists of an application of the basic sampling procedure outlined in Sect. 7.3.1, but limiting the candidate sample to either households containing children or large households. For the simulation of households containing children, the candidate sample is limited to those households alone. For the simulation of large households, a household size constraint is specified, and the selection process is carried out for candidate samples limited to each household size in descending order (i.e. a random sample of households of eight persons or more are considered first, followed by those of seven persons, etc.) The household size constraint is dropped after this stage. The order which yields the most accurate simulation within these two stages is determined by a Monte Carlo process of repeated sampling (see Farrell et al. 2012 for a full discussion).

Configuration 2 improves the level of convergence for EDs which may have characteristics proportionally disparate to national population distributions. This is carried out by prioritising those households containing individuals who may be able to satisfy ED-level population distributions which are far greater than the national-level distribution. For example, if the ED share of 15–20-year-old males exceeds that of the national population by a predefined threshold, households containing individuals of this constraint classification will be simulated first. The procedure for determining this threshold is based on an accuracy vs. efficiency trade-off, discussed in Farrell et al. (2012). The degree of proportional disparity is calculated for each constraint, with a ranking in descending order of disparity in place if a number of constraints are greater than the predefined threshold. Households are then assigned a weight according to their ability to fill the quota for the constraint of greatest disparity. Those households with the greatest weight are sorted randomly and considered for synthesis using the quota sampling synthesis procedure. This allows for greater convergence for EDs with particular population patterns that are different to national distributions, capturing a greater degree of spatial heterogeneity (for a full discussion of the process, see Farrell et al. 2012).

Configuration 4 is carried out after configuration 3 to counteract the restrictions imposed by diminishing quotas. This involves the broadening of constraints and is carried out as follows. If the algorithm fails to assign any further households due to overly restrictive quotas, one constraint is removed. This increases the search space allowing households to be considered that were once excluded. This is repeated one constraint at a time until either all remaining quota counts are filled or all constraints have been removed. If the algorithm fails to assign an adequate number of individuals during this procedure, individuals are assigned at random to meet the required population. Constraints are removed in reverse order of the degree to which they influence household income, determined by pre-synthesis regression analysis (see O'Donoghue et al. 2012 and Farrell et al. 2012) for a full outline of this procedure). This design minimises subjectivity, whereby the broadening of constraints is only introduced when absolutely necessary and in a fashion which ensures that those variables that explain the greatest level of variability are retained to the greatest extent. Sometimes all quotas are filled and this stage is skipped.

It may be suggested that broadening the constraints in such a manner may cause validation issues to arise in that the distribution for larger households or under-represented groups may be less robust. To ensure this does not occur, validation of the QS output is an integral component of the model's construction. Section 7.4 outlines the validation methods used within the SMILE model.

7.4 Validation of the New Created Dataset

Once the base dataset has been synthesised, validation is carried out to ensure the simulated populations are consistent with empirical benchmarks both internal and external to the simulation process. This is difficult, as the creation of synthetic microdata is motivated by non-existence of such data for small geographic areas. However, as Oketch and Carrick (2005) point out, it is only through validation that the credibility and reliability of a microsimulation model, and thus the regional welfare distributions in SMILE, can be assured. Caldwell (1996) provides an overview of validation techniques that may be used to ensure the robustness of synthetically created data. For SMILE, two validation procedures are employed:

- In-sample validation to determine whether the spatial relationship of overlapping variables has been maintained
- Out-of-sample validation to determine whether simulated data represent the spatial distribution of non-constrained variables

7.4.1 In-Sample Validation

In-sample validation aggregates simulated microdata for comparison with the regional benchmarks used to constrain the simulation and thus provides the primary method by which the statistical matching procedure is appraised. In doing so, this procedure also ensures the correct spatial distribution of the primary determinants of household welfare. For quota sampling, the in-sample procedure employed compares the proportional correlation of each constraint variable to those in the SAPS. In our experience, the correlation coefficient places a greater weight on the size of the district rather than on the distribution of ages when using absolute totals, and thus validation according to proportional correlations is preferred.

7.4.1.1 In-Sample Validation: Results

For each constraint variable, proportional correlations were found to be close to 1, with almost all being greater than 0.98, and the majority being greater than 0.99 (O'Donoghue et al. 2012). The apportioned sampling procedure has resulted in an equal distribution of fit for both children and adults, thus counteracting imbalances quoted previously. It was found, however, that 15–25 age groups displayed a degree of fit that was on average less than that of other age groups. Upon closer inspection, this seems to relate to regions with a great number of young people relative to the number of adults, indicating those living either in institutions (primarily boarding schools, university residences or shared student apartments) which are under-reported in the survey. Although the ordering procedure has remedied this issue somewhat, the sampling algorithm still struggles to find enough young individuals living alone to produce enough young people in certain districts dominated by students. As this issue affects only a relatively small number of districts and because of data limitations, our intention is return to this issue later when a more concentrated analysis in relation to education is required as in the case of Wu et al. (2008).

7.4.2 Out-of-Sample Validation

To complement our validation of constrained variables, we validate the non-constrained variable of disposable income. This is carried out by comparing SMILE county-level aggregates to county-level poverty statistics of the National Survey of Household Quality (NSHQ) from 2001/2, reported in Watson et al. (2005). Although a survey primarily aimed at analysing housing issues, the NSHQ collects data on disposable income and is representative at the county level. In doing so, poverty rates are expressed as a function of the national average in SMILE and compared to the relative poverty headcount in the NSHQ at levels of 50% and 60% of median disposable income.

7.4.2.1 Out-of-Sample Validation: Results

The observed relationship between the two data sources is high, with a correlation of 0.78 and 0.79 at the 50% and 60% levels, respectively (for a full description, see O'Donoghue et al. 2012). This indicates that the ranks for all the areas are good, but there is much greater spread in the NSHQ than in the SMILE output. The reason for these differences is due to additional spatial heterogeneity in incomes that is not captured by our constraint variables. This is confirmed by O'Donoghue et al. (2012), whereby controlling for region improved model fit in regressions to predict male earnings.

Suggested solutions have included the use of alternative or additional constraints, sampling of micro-units from the same (aggregated) spatial area (Voas and Williamson 2000), separate matching methods for different spatial clusters (Smith et al. 2009) or alternative sets of constraints depending on the purpose (Chin and Harding 2006). Given the disproportionate computational cost of employing additional constraints reported by Miller (2001), along with the desire to have a single model configuration for all uses, we choose instead to adopt a calibration procedure.

7.5 Calibration

7.5.1 Procedure Overview

The calibration procedure to capture additional spatial heterogeneity draws on methodologies employed in a number of fields including dynamic inter-temporal simulation (O'Donoghue 2010), macro–micro literature to capture the impact of macroeconomic changes on income distribution (Ahmed and O' Donoghue 2007, 2008), data synthesis under IPF (Ballas et al. 2006) and development economics literature focussing on inter-country heterogeneity (Bourguignon et al. 2002).

The purpose of the calibration procedure is to align disaggregated data within SMILE to exogenous spatial distributions of income. The procedure presented here and discussed in full in Morrissey and O'Donoghue (2011) operates in two stages: equations determining the presence of an income are first estimated, followed by those predicting the level of that income. This process is described below.

A set of nested choice equations, pertaining to labour market characteristics and the presence of other market income sources, are estimated for each individual. Each equation predicts each individual's probability of having a certain labour market characteristic y_i depending on their set of explanatory factors X_i and estimated parameters β_i. Depending on the format of the dependant variable in question, binary choice (logit), multiple choice (reduced form multinomial logit) and logged income regression models are used. These take the general form

$$y_i^* = g(\beta_i X_i) + \varepsilon_i, \quad (7.1)$$

where $g(\beta_i X_i)$ represents the deterministic or explained elements which affect the probability of having the characteristic y_i, and ε_i constitutes the unexplained residual satisfying the condition $E[\varepsilon_i|x_i]=0$.

This model simulates the labour market variable $y = 1$ if $y_i^* > 0$. A decision rule is created to determine which individuals' characteristics will be changed to meet the exogenous small area totals. For each individual, we rank our predicted variable y_i^* defined in (7.1) such that we select the N cases simulated with the highest value of y_i^*. For binary variables our calibration routine requires N cases of a particular unconstrained variable in the relevant district. In multiple choice models, a similar method is developed, ranking y_i^* for each choice j in turn to be consistent with externally defined N_j. Use of the residual component value ε_i is important as using the deterministic component alone excludes consideration of those with a low probability of an event occurring (Morrissey and O'Donoghue 2011). For example, lone parents tend to have a low probability of working. Without considering the residual component, alignment will rank y_i and tend to produce no lone parents in work. However, even if the in-work probability is low, there are some who work. The inclusion of the error term in the ranking will tend to shuffle the data so that some of those with low risk will be predicted to work. These models are estimated on the original LII micro-dataset and then simulated consecutively for each ED according to the distribution of X_i characteristics of units in the synthetic spatial dataset.

The spatial distribution of unconstrained labour market characteristics and other market income sources are calibrated against regional SAPS totals. Once the correct distribution of these variables has been established, the level of income is calibrated according to external county-level national accounts. Definitional differences prohibit absolute adjustment in the calibration of income, as scaling average income by source to the national accounts total can affect the distributional properties of the data. Thus, the calibration procedure is augmented to ensure average county income by income source corresponds to county-level national accounts. To overcome this, the ratio of average income by source is scaled proportionally to the national average. This allows the same distribution properties of the underlying income data to be largely maintained.

7.5.2 Calibration Results

The correlation between the unconstrained variables of employment status (in-work vs. out-of-work, employee, unemployed, retired), occupation and industry are calculated before and after calibration. It was found that there was substantial variability in correlation results across counties. Correlations vary from highs of greater than 0.8 to a very poor, almost random, relationship close to zero. Furthermore, we also see that the correlations decrease by layer of hierarchy whereby higher-order

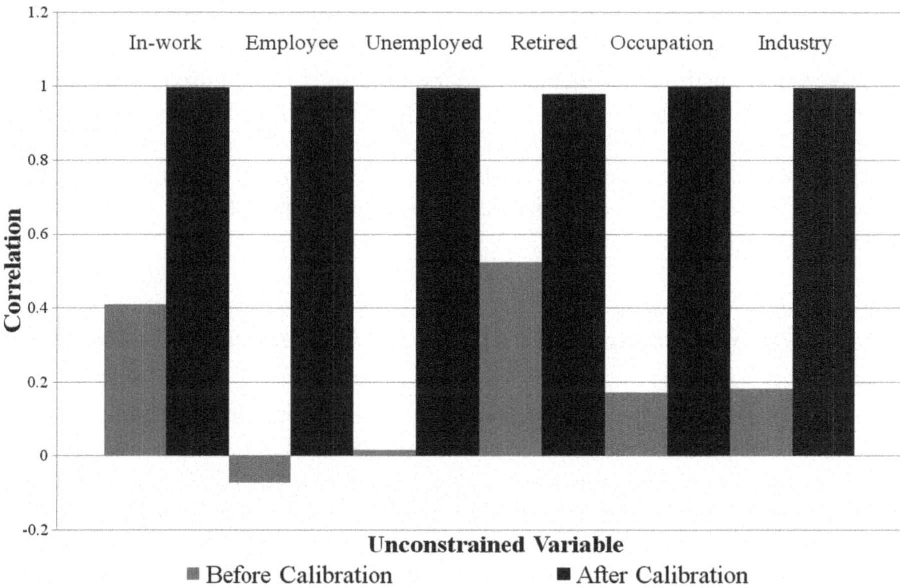

Fig. 7.2 Average unconstrained variable correlation (males)

characteristics such as in-work or retired perform better than lower-order characteristics such as employee or industry.

Post-calibration, a clear improvement in correlations is found, with all correlations close to 1. Excluding retirees, almost all variables display a correlation in excess of .99, with those retired being marginally less at a rate of .938–.998. This is still within bounds of acceptability, especially as those retired constitute a smaller proportion of the population than other labour market variables, and thus a small absolute discrepancy results in a greater proportional difference such as this. As a result, it may be concluded that the calibration procedure is effective at ensuring that the simulated population is consistent with known spatial distributions. Based on results reported in O'Donoghue et al. (2012), county-level correlations with SAPS-defined target values are averaged in Fig. 7.2, illustrating the degree of improvement, post-calibration.

7.6 An Application of SMILE: Spatial Analysis of Income Inequality and Redistribution in Ireland

Having outlined the procedure of data synthesis and calibration, the data created by SMILE is now applied to analyse the spatial distribution of welfare and income redistribution in Ireland. As indicated in Sect. 7.5, the calibration procedure aligns income data produced by SMILE to 2005 National Accounts county-level distributions of market-level income.[2] However, an accurate measure of poverty at any spatial level must take into account the social welfare transfers within a country. Thus, these newly aligned market income data are used to calculate disposable income (income net of taxes and benefits) for each individual in Ireland using SMILE's tax-benefit component. This section discusses the results of that process whilst also analysing the ability of the Irish tax-benefit system to reduce income inequality within and between spatial entities (see O'Donoghue et al. 2012, for a full description of this process).

7.6.1 The Spatial Incidence of Income Redistribution in Ireland

Figure 7.3 presents the spatial distribution of the ratio of tax to disposable income, whilst Fig. 7.4 presents the spatial distribution of benefits to disposable income. These figures present their respective spatial distributions at the ED level for Ireland and are created using the data produced by SMILE (O'Donoghue et al. 2012). Upon examination, Fig. 7.3 clearly highlights that the EDs with the lowest ratios of taxes to disposable income are located in the border area, parts of the midlands and the south west (darker shading). The EDs that pay the highest ratio of taxes to disposable income are located on the eastern sea board (light shading). It should be noted that the areas with the highest ratios of income taxes to disposable income correspond to the areas with the highest levels of market income.

Figure 7.4 displays the ratio of benefits to disposable income. One can see that the distribution of benefits to disposable income is more concentrated, with the most rural districts having the highest concentration of benefits to disposable income (darker shading). This is due to a greater number of benefit recipients residing in these districts, as benefits are largely at flat rate. In particular, it is found that the proportion of individuals of pension age in an ED is a key determinant of the concentration of benefits to disposable income.

7.6.2 Impact of Tax-Benefit Policy on Between and Within Group Income Inequality

In addition to identifying the spatial incidence of income redistribution, we would also like to understand how income is redistributed within and between spatial entities. To do this, we examine the variability of incomes between individuals within and between regions by aggregating a measure of individual income inequality into population subgroups. In this way, one can decompose total variability of incomes into a factor attributed to between group variability across space and variability within a district (within group variability).

[2] Market income is income before the deduction of income taxes and addition of benefits.

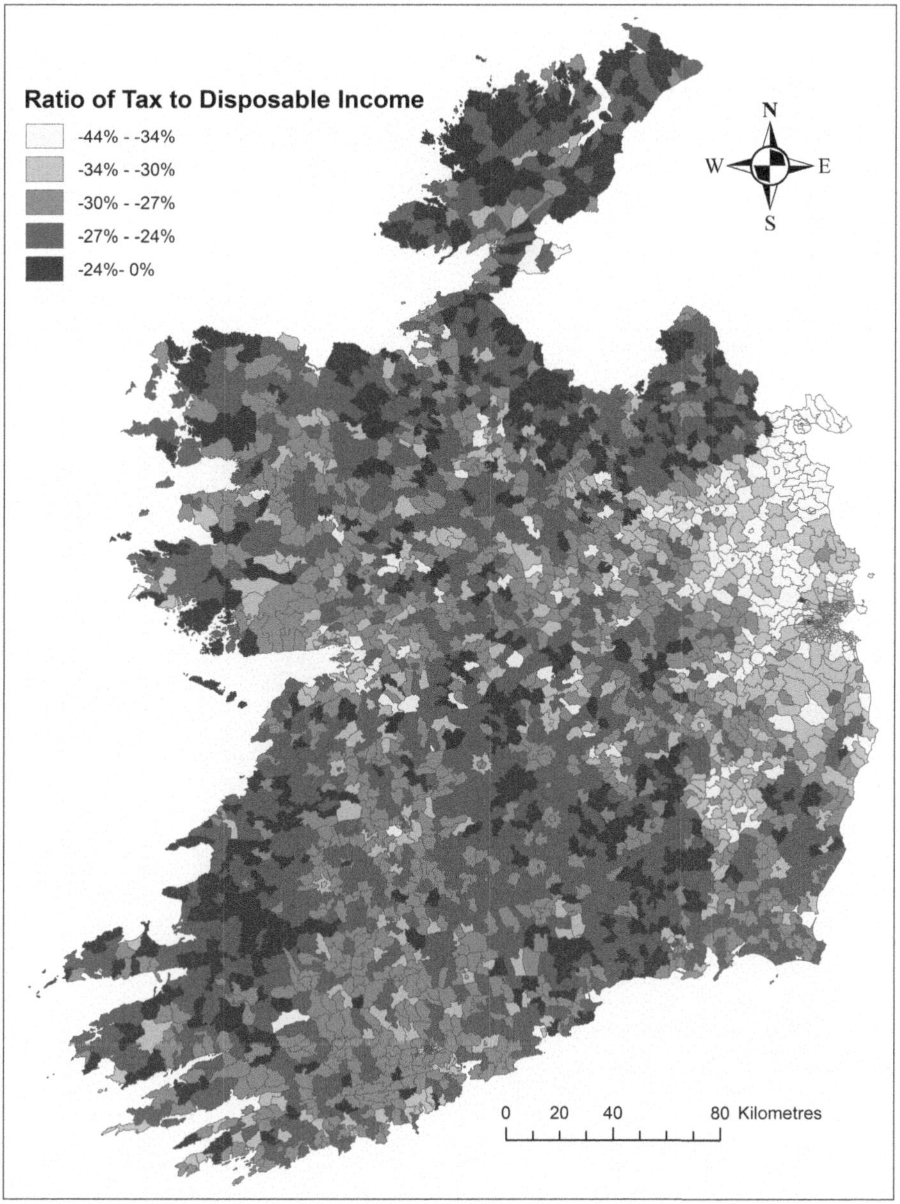

Fig. 7.3 Distribution of taxes as a percentage of disposable income for EDs in Ireland

The index used for calculating within group variability (I_w) uses the I_2 index, an index for calculating the degree of economic inequality. The I_2 index may be defined as half the squared coefficient of variation, $\left(\frac{\sigma^2}{2\mu^2}\right)$ (see Jenkins 1995), where σ signifies the standard deviation of incomes and μ, the mean population lifetime income.

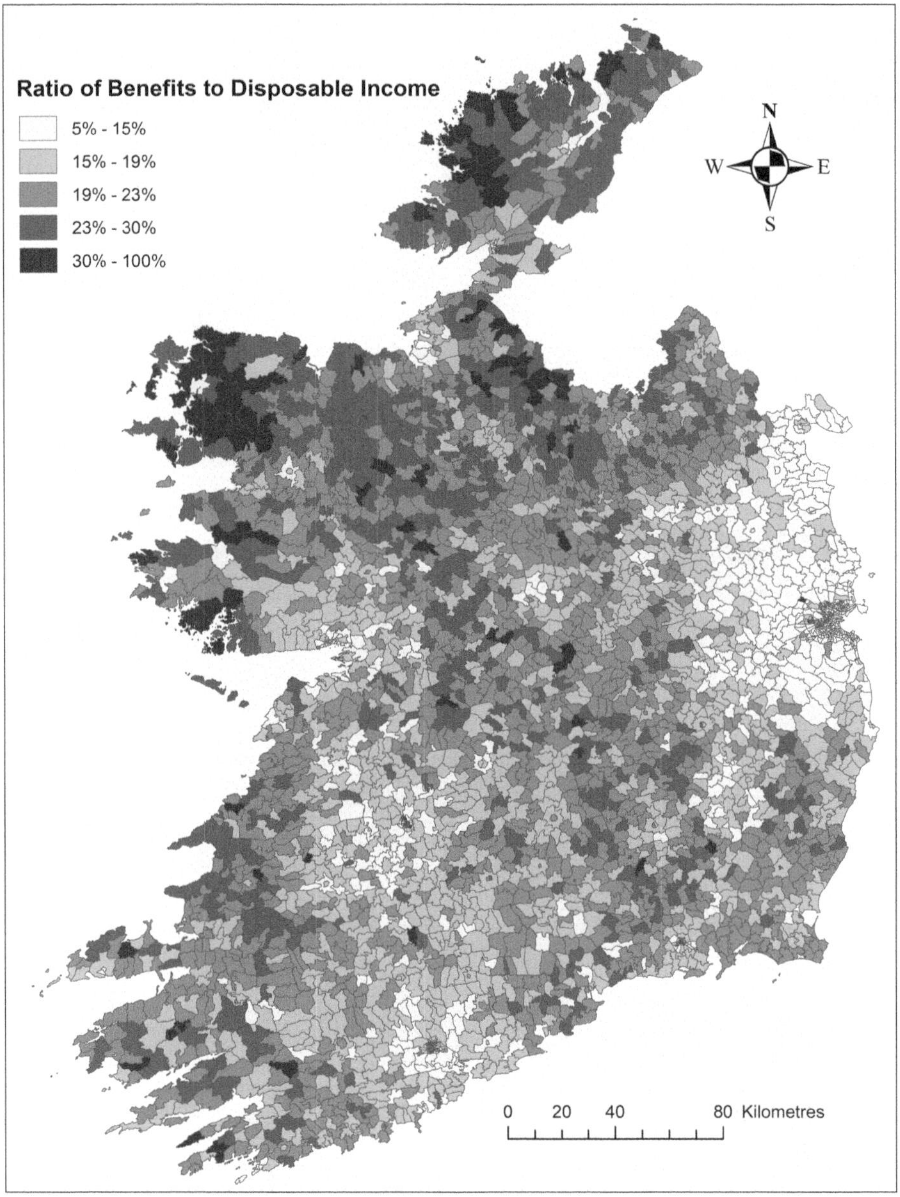

Fig. 7.4 Distribution of benefits as a percentage of disposable income for EDs in Ireland

Thus, I_w may be defined as

$$I_w = \sum_j w_j I_j, \tag{7.2}$$

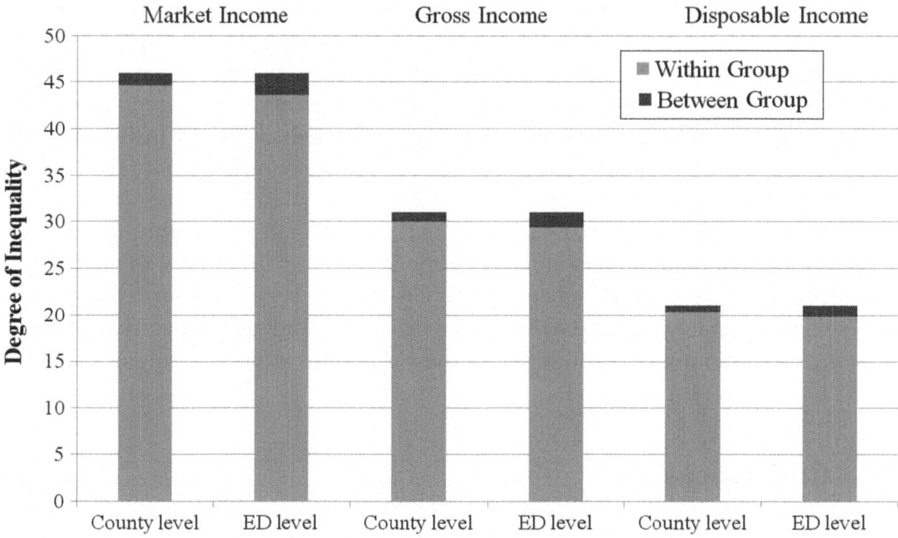

Fig. 7.5 Comparison of county and ED-level between group income inequalities

where $w_j = v_j^2 f_j^{-1}$, v_j is the income share of each person j in a given spatial group and f_j is the population share of person j in a given spatial group, in this case $(1/n)$. I_j stands for the inequality measure, I_2.

The index for between group variability (I_b) is defined as

$$I_b(y) = \frac{1}{2}\left[\frac{1}{n}\sum_j \left(\frac{\mu_j}{\mu}\right)^2 - 1\right], \tag{7.3}$$

where μ_j is the mean lifetime income for person j and μ, the mean population lifetime income.

Based on analysis in O'Donoghue et al. (2012), Fig. 7.5 displays the between and within group inequality by income at the ED level. Income is displayed according to three different definitions: market income (income before adding benefits and deducting income taxes), gross income (income after adding benefits but before deducting income taxes) and disposable income (income after adding benefits and deducting income taxes). Each definition of income displays a different stage of the tax-benefit process and allows for the effect of each stage of the tax-benefit process on spatial inequality to be elicited.

Figure 7.5 illustrates that between district inequality accounts for a very small proportion of overall inequality, with most inequality existing within districts (between families). It may also be noted that the share of within group inequality is marginally greater at the more aggregate spatial level of county than at the ED level.

One can see that the overall level of inequality reduces as one adds benefits and subtracts taxes to get gross income and disposable income, respectively. This has an insignificant effect on between group inequality, however, as the proportion accounted for by between group inequality remains roughly the same. Thus, one may conclude that tax-benefit policy does not act to reduce spatial inequality (i.e. between spatial group), but rather it acts more to reduce between family (i.e. within spatial group) inequalities.

Thus, in the absence of pre-existing spatial microdata, it is only through the use of spatial microsimulation techniques that these spatial distributions and measures of inequality may be elicited and compared. Given that a regional approach to policy analysis has potential to improve welfare disparities, such distributional data may provide input into determining areas of prioritisation for future spatial targeting of policy.

7.7 Conclusions

Lack of spatial microdata has significantly limited spatial analyses of welfare in Ireland. This chapter outlines the creation of an Irish spatial microsimulation model to overcome this and illustrates how within- and between-region welfare analyses at the small area ED level may be achieved.

As the household has been deemed the most appropriate unit of micro-level welfare analysis, a greater level of complexity is imposed on the choice of simulation process. The means by which SMILE has accommodated this requirement has evolved as successive versions have been developed. Initially, IPF was employed (Ballas et al. 2006), but a desire to employ actual microdata motivated the use of SA procedures in the next version of the model (Morrissey et al. 2008). SA, however, is computationally intensive and thus precludes the use of repeated syntheses or development of future projections. As a result, the development of the current version of SMILE has involved the creation of a computationally efficient method known as quota sampling (Farrell et al. 2012). The conceptual and practical implications of this procedure have been outlined in this chapter.

As with all spatial microsimulation models, the credibility of results relies on how well actual population distributions are emulated. In order to ensure reliability of estimated welfare distributions, extensive validation procedures are required. The performance of quota sampling has been assessed using both in-sample and out-of-sample validation. Whilst these validation results are quite good given that different datasets were used, we note in particular an issue in relation to unexplained spatial heterogeneity. This has prompted a calibration procedure. This is carried out in two steps whereby an accurate distribution of labour force variables is simulated, followed by an alignment procedure whereby market incomes are readjusted to be representative of national accounts. On completion of the alignment process, SMILE offers a fully representative profile of labour force participation and market incomes at both the household and small area level. In the absence of actual small area microdata, calibration ensures the most reliable estimation of spatially referenced microdata.

Using this model, the spatial distribution of income and the impact that the tax-benefit system has on changing the income distribution has been estimated. It was found that disposable income is on average lower in rural than urban areas with transfers from urban to rural areas. These results correspond to those of Morgenroth (2010) who developed an analysis of the regional transfers across Ireland and showed that there is a transfer of resources from the Greater Dublin Area and southwest regions of the country to the rest of the country. Furthermore, it has been found that Irish tax-benefit policy is more effective in reducing within-region inequality than between-region inequality (O'Donoghue et al. 2012).

As such, this chapter illustrates the creation of a spatial profile of disposable income and welfare redistribution in Ireland using spatial microsimulation techniques. In doing so, individual-level, spatially referenced data have facilitated distributional analyses which would otherwise have been infeasible. Such analyses can deepen our understanding of the determinants of inequality and poverty and lead to improvements in the design of policies tailored to local conditions.

References

Ahmed, V., & O' Donoghue, C. (2007). *CGE-microsimulation modelling: A survey* (MPRA Paper 9307). University Library of Munich, Germany.

Ahmed, V., & O' Donoghue, C. (2008). *Welfare impact of external balance in Pakistan: CGE-microsimulation analysis* (MPRA Paper 9267). University Library of Munich, Germany.

Ballas, D., Clarke, G. P., Dorling, D., Eyre, H., Rossiter, D., & Thomas, B. (2005a). SimBritain: A spatial microsimulation approach to population dynamics. *Population, Space and Place, 11*(1), 13–34.

Ballas, D., Rossiter, D., Thomas, B., Clarke, G., & Dorling, D. (2005b). *Geography matters: Simulating the local impacts of national social policies* (Joseph Rowntree Foundation contemporary research issues). York: Joseph Rowntree Foundation.

Ballas, D., Clarke, G., & Wiemers, E. (2006). Spatial microsimulation for rural policy analysis in Ireland: The implications of CAP reforms for the national spatial strategy. *Journal of Rural Studies, 22*(3), 367–378.

Bell, P. (2000). *GREGWT and TABLE macros – Users guide*. Canberra: Australian Bureau of Statistics. Unpublished.

Bourguignon, F., da Silva, P. L., & Stern, N. (2002). *Evaluating the poverty impact of economic policies: Some analytical challenges* (Technical report). Washington, DC: The World Bank.

Caldwell, S. (1996). Health, wealth, pensions and life paths: The CORSIM dynamic microsimulation model. In A. Harding (Ed.), *Microsimulation and public policy*. Amsterdam: North Holland.

Chin, S-F., & Harding, A. (2006). *Regional dimensions: Creating synthetic small-area microdata and spatial microsimulation models* (National Centre for Social and Economic Modelling, Technical Paper no. 33). https://guard.canberra.edu.au/natsem/index.php?mode=download&file_id=648. Accessed 18 Sept 2011.

Chin, S.-F., Harding, A., Lloyd, R., McNamara, J., Phillips, B., & Vu, Q. N. (2005). Spatial microsimulation using synthetic small-area estimates of income, tax and social security benefits. *The Australasian Journal of Regional Studies, 11*(3), 303–335.

Edwards, K. L., & Clarke, G. P. (2009). The design and validation of a spatial microsimulation model of obesogenic environments for children in Leeds: SimObesity. *Social Science and Medicine, 69*(7), 1127–1134.

Elbers, C., Fujii, T., Lanjouw, P., Özler, B., & Yin, W. (2007). Poverty alleviation through geographic targeting: How much does disaggregation help? *Journal of Development Economics, 83*(1), 198–213.

Farrell, N., O'Donoghue, C., & Morrissey, K. (2012). *Spatial microsimulation using quota sampling*. Teagasc Rural Economy Development Programme Working Paper Available online: http://www.agresearch.teagasc.ie/rerc/workingpapers.asp. Accessed on 7 July 2012.

Haase, T., & Foley, R. (2009). *Feasibility study for a local poverty index*. Dublin: Combat Poverty Agency.

Hajnal, Z. L. (1995). The nature of concentrated urban poverty in Canada and the United States. *The Canadian Journal of Sociology/Cahiers canadiens de sociologie, 20*(4), 497–528.

Hynes, S., Morrissey, K., O'Donoghue, C., & Clarke, G. (2009). Building a static farm level spatial microsimulation model for rural development and agricultural policy analysis in Ireland. *International Journal of Agricultural Resources, Governance and Ecology, 8*(2), 282–299.

Jencks, C., & Mayer, S. E. (1990). The social consequences of growing up in a poor neighborhood. In L. Lynn & M. McGeary (Eds.), *Inner-city poverty in the United States* (pp. 111–186). Washington, DC: National Academy Press.

Jenkins, S. P. (1995). Accounting for inequality trends: Decomposition analyses for the UK, 1971–86. *Economica, 62*(245), 29–63.

Miller, E. J. (2001). *The Greater Toronto area travel demand modelling system version 2.0, Volume I: Model overview*. Toronto: Joint Program in Transportation, University of Toronto.

Morgenroth, E. (2010). Regional dimension of taxes and public expenditure in Ireland. *Regional Studies, 44*(6), 777–789.

Morrissey, K., Clarke, G., Ballas, D., Hynes, S., & O'Donoghue, C. (2008). Examining access to GP services in rural Ireland using microsimulation analysis. *Area, 40*(3), 354–364.

Morrissey, K. and O'Donoghue, C. (2011). The spatial distribution of labour force participation and market earnings at the sub-national level in Ireland. *Review of Economic Analysis, 3*, 80–101.

Norman, P. (1999). *Putting iterative proportional fitting (IPF) on the researcher's desk*. (Working Paper 99/03 ed.). School of Geography, University of Leeds, Leeds.

O'Donoghue, C. (2010). *Life-cycle income analysis modelling*. Saarbucken: Lambert Academic Publishing AG & Co.KG.

O'Donoghue, C., Hynes, S., Morrissey, K., Ballas, D., & Clarke, G. (Eds). (2012). Spatial microsimulation for rural policy analysis. Dordrecht: Springer.

O'Leary, E. (2003). Aggregate and sectoral convergence among Irish regions: The role of structural change, 1960–96. *International Regional Science Review, 26*(4), 483–501.

Oketch, T., & Carrick, M. (2005, January). *Calibration and validation of a micro-simulation model in network analysis*. Proceedings of the 84th TRB Annual Meeting, Washington, DC.

Ravallion, M., & Jalan, J. (1997). *Spatial poverty traps?* (World Bank Policy Research Working Paper No. 1862). Available at SSRN: http://ssrn.com/abstract=597203. Accessed 15 Aug 2011.

Smith, D. M., Clarke, G. P., & Harland, K. (2009). Improving the synthetic data generation process in spatial microsimulation models. *Environment and Planning A, 41*(5), 1251–1268.

Tanton, R., Williamson, P., & Harding, A. (2007, August). *Comparing two methods of reweighting a survey file to small area data: Generalised regression and combinatorial optimisation*. 1st Gen. Conf. International Microsimulation Association, Vienna.

Tanton, R., McNamara, J., Harding, A., & Morrison, T. (2009a). Rich suburbs, poor suburbs? Small area poverty estimates for Australia's eastern seaboard in 2006. In A. Zaidi, A. Harding, & P. Williamson (Eds.), *New frontiers in microsimulation modelling*. London: Ashgate.

Tanton, R., Vidyattama, Y., McNamara, J., Vu, Q., & Harding, A. (2009b). Old, single and poor: Using microsimulation and microdata to analyse poverty and the impact of policy change among older Australians. *Economic Papers, 28*(2), 102–120.

Vidyattama, Y., Cassells, R., Harding, A., & Mcnamara, J. (2011). Rich or poor in retirement? A small area analysis of Australian private superannuation savings in 2006 using spatial microsimulation. *Regional Studies*. doi: 10.1080/00343404.2011.589829. Available online: http://www.tandfonline.com/doi/abs/10.1080/00343404.2011.589829. Accessed 25 Aug 2011.

Voas, D., & Williamson, P. (2000). An evaluation of the combinatorial optimisation approach to the creation of synthetic microdata. *International Journal of Population Geography, 6*, 349–366.
Watson, D., Whelan, C. T., Willams, J., & Blackwell, S. (2005). *Mapping poverty: National regional and county patterns* (Combat poverty agency research series, Vol. 34). Dublin: Combat Poverty Agency.
Williamson, P. (2009, April 9). *Creating synthetic sub-regional baseline populations*. Paper presented to ESRC Microsimulation Series, London.
Wong, D. W. S. (1992). The reliability of using the iterative proportional fitting procedure. *The Professional Geographer, 44*(3), 340–348.
Wu, L., & Wang, Y. (1998). An introduction to simulated annealing algorithms for computation of economic equilibrium. *Computational Economics, 12*, 151–169.
Wu, B., Birkin, M., & Rees, P. (2008). A spatial microsimulation model with student agents. *Computers Environment and Urban Systems, 32*, 440–453.

Chapter 8
Linking Static Spatial Microsimulation Modelling to Meso-scale Models: The Relationship Between Access to GP Services and Long-Term Illness

Karyn Morrissey, Graham Clarke, and Cathal O'Donoghue

8.1 Introduction

There are a wide range of methodological frameworks and techniques for policy evaluation and socio-economic impact assessment. For example, models built on aggregate datasets (such as the census and national-level surveys) are widespread and have proved very fruitful in many areas of policy analysis (see, e.g. Longley and Batty 2001; Stillwell and Clarke 2004). Nevertheless, the complex dynamics which underlie health-care markets, as emphasised by Sassi and Hurst (2008), call for more sophisticated tools to help in the formulation and evaluation of appropriate and effective health policies. In order to formulate such policies, it is necessary not only to understand the nature and the operation of the health sector at a macro-level but also to evaluate the likely impact of these policies on health activity at the local level. In particular, there is a need to understand, estimate or predict which individuals (given their demographic and socio-economic characteristics) and areas are most likely to benefit from a change in health-care policy.

Thus, policy-relevant modelling is a challenging research area which is better suited to a modelling framework which emphasises individual-level processes at the local level while encompassing aggregated process, such as service provision at the macro-level. Spatial microsimulation modelling is a means of synthetically creating geographically referenced micro-data. As pointed out by Ballas and Clarke (2001), individual-level issues may be usefully addressed in a spatial microsimulation

K. Morrissey (✉)
School of Environmental Sciences, University of Liverpool, Liverpool, UK
e-mail: karyn.morrissey@liv.ac.uk

G. Clarke
School of Geography, University of Leeds, Leeds, UK

C. O'Donoghue
Rural Economic Development Programme, Teagasc, Ireland

R. Tanton and K.L. Edwards (eds.), *Spatial Microsimulation: A Reference Guide for Users*, Understanding Population Trends and Processes 6, DOI 10.1007/978-94-007-4623-7_8,

framework – as also demonstrated by the research by Lymer et al. on aged care needs and disability (Lymer et al. 2009, 2008). However, to ensure that meso-interactions are accounted for, the application of spatial microsimulation models to health research requires a model component that accounts for these interactions. Within this context, this chapter aims to improve the applicability of the data created by a microsimulation model, not by increasing the complexity of the simulation framework, but by integrating the newly created data within a meso-level, spatial interaction framework. Such an analysis provides policymakers with information on both demand for and supply of health-care services, thus allowing them to shape future service provision and target existing health resources in a more efficient and effective manner.

Previous work within the health-care and service provision literature linking spatial microsimulation models and spatial interaction models include work by Morrissey et al. (2008) which used a spatial interaction model (SIM) to examine whether the current spatial distribution of GPs in County Galway matched demand for these services. Morrissey et al. (2010) also examined the spatial distribution of depression at the small area level in Ireland and levels of access to both acute and community psychiatric care. Tomintz et al. (2008) linked a spatial microsimulation model to a location-allocation model to optimally locate smoking cessation clinics in Leeds, Yorkshire. Thus, linking spatial microsimulation models to macro-level models, such as SIM, provides a powerful tool for examining a wide range of policy questions (Smith et al. 2009). Given previous work in this area, the key innovation offered in this chapter is the ability to endogenise the ease of access to GP services within a statistical model of long-term illness (LTI). This is achieved by combining the output from a SIM within the dataset produced by a spatial microsimulation model.

8.1.1 Modelling Health Status

The last five decades have seen a dramatic improvement in the health and longevity of people in countries within the Organisation for Economic Cooperation and Development (OECD) (Sassi and Hurst 2008). However, at the same time, non-communicable diseases are currently the main cause of both disability and mortality worldwide, with the burden of long-term chronic illness proportionally larger in OECD countries (WHO 2000). At the individual level, widespread health disparities among population groups are becoming more pronounced (Mackenbach 2006; Sassi and Hurst 2008). Research in a number of countries has found that individuals in lower socio-economic and income categories have worse health than individuals in higher socio-economic and income categories, with a continuous gradient observed between the two extremes (Safaei 2007). In Ireland, previous research on the determinants of ill-health has found that medical card ownership (used as a proxy for being economically disadvantaged) is a consistent indicator of poor health for the Irish population (Kelleher et al. 2002). Tay et al. (2004) found that (self-assessed) low financial security and dissatisfaction with work were strong indicators of ill-health. Madden (2008) found that income poverty was a good, though not a perfect, indicator of health ‘poverty’ in Ireland.

At a disaggregated spatial scale, the existence of area-based differences in individual health status has long been established in British literature (Mitchell et al. 2002). Since the Black Report in 1980, a number of important studies have confirmed that spatial health inequalities exist on a wide (and widening) scale, not just in Britain, but across the globe (Wilkinson and Pickett 2006; Dorling et al. 2007; Shaw et al. 2008). These studies found that clusterings of similar individual-level factors such as income level, age and employment status increase the risk of mortality and specific morbidities across space. This research, however, has focused on endogenous individual-level factors that affect health status. Exogenous forces, such as health service provision, medical prices and the education system within a country may have an indirect effect on individual health.

With regard to service provision, the centralised provision of national health policy and the costs associated with providing health services in sparsely populated areas (Asthana et al. 2002) mean that health-care services are not provided evenly across space. Research in the UK has focused on the effect of physical access on the likelihood of obtaining medical treatment. For example, Jones et al. (2010) and Campbell et al. (2000) found that diagnosis of cancer on death, when coupled with social disadvantages, may be associated with poorer geographical access to health services care. With regard to access to GP services and cancer survival rates, research in Northern England found that late-stage presentations in breast and colorectal cancer patients, and poorer survival in prostate cancer patients, were associated with longer car journeys to GP surgeries (Jones et al. 2008).

However, from these studies, it is not possible to assert that poorer access and service provision will adversely affect the health status of the residents of an area. Given the strong (and increasingly important) role primary care plays in the management of long-term illness ('LTI') (Department of Health and Children 2001), this chapter examines the relationship between LTI and GP service provision at the small area level. Thus, building on previous UK-based research on the relationship between treatment and access, this chapter continues the research and endogenises access to GP services as a determinant of LTI. We do this by combining the output from a spatial microsimulation model within a spatial interaction framework.

8.2 Methodology

8.2.1 Step 1: The Creation of Spatially Disaggregated Data Using a Spatial Microsimulation Model

Numerous attempts have been made in recent years to conceptualise the role and reciprocal influences of different groups of health determinants (Sassi and Hurst 2008). Quantitative models help to understand the pathways and determinants of health status by attempting to capture and quantify the effects of individual health determinants and the interdependencies between these factors. However, to establish the key determinants of health status (or in this case LTI), a large variety of data is required at the individual level (which is not often publically available even if it actually exists). The first step in our modelling framework is therefore to create spatially disaggregated micro-data containing the necessary variables to estimate the determinants of LTI.

As outlined by Morrissey et al. (2008), although there are a number of datasets containing health data for Ireland, for example, the Living in Ireland (LII) survey, the spatial identifiers in these datasets tend to be at a very high level of spatial aggregation. The LII contains two location variables, a NUTS-3 regional variable (covering eight regions) and a 12-category locational variable, categorised into the five cities in Ireland, a category for Dublin County, an 'open-countryside' category and five categories for towns of varying sizes. On the other hand, the Irish Small Area Population Statistics (SAPS) contains detailed demographic and socio-economic data at the small area level, electoral district (ED) level, county level and regional level, but it does not contain any health variables. EDs are the smallest geographical output area for all statistics produced in Ireland. There are 238 EDs in Co. Galway. The population in any one ED ranges from a low of 77 individuals to a high of 8,629, with an average across all EDs of 719. By merging the relevant health data from the LII survey with the ED census data (i.e. the SAPS dataset), a much richer micro-level dataset can be created. Spatial MSM may be used to accomplish this.

A version of SMILE is the static spatial microsimulation model used for this work (Morrissey et al. 2008, 2010; see also Chap. 7 of this book for a description of a later version of SMILE that uses quota sampling). It uses a combinational optimisation technique, simulated annealing, to match the LII (2000) and SAPS (2002) datasets, matching on age, sex, household size and education level, creating a micro-level synthetic dataset for the whole population of Ireland, which includes health variables. For a full discussion on the algorithm and datasets used to create the statistical match, please see Morrissey et al. (2008). Ballas et al. (2006) also provide an outline of the simulated annealing process and various other methods that may be used to create synthetic data at varying spatial scales. Figure 8.1 represents the inputs, modelling process and outputs of the creation of the spatially disaggregated data.

The dataset created by the SMILE model contains demographic, socio-economic, labour force, income and, importantly, health variables for both individuals and family units. The health data created by SMILE has previously been used to examine differentials in GP utilisation rates between urban and rural areas in County Galway (Morrissey et al. 2008) and access acute hospitals for individuals suffering from depression in Ireland (Morrissey et al. 2010). Table 8.1 contains the most important health status drivers found in the matched SMILE dataset. It is important to note that the LTI variable is a self-reported variable. Individuals were asked if they suffered from a variety of different physical and emotional conditions and responded yes or no. Although SMILE produces a geo-referenced population dataset for the whole of Ireland, for the purposes of this chapter, only data for County Galway will be used. Figure 8.2 presents the population within the study area, Co. Galway, and its location within Ireland.

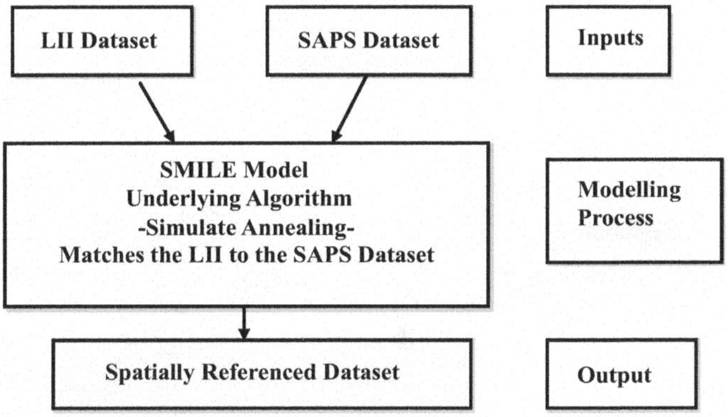

Fig. 8.1 SMILE: Spatial microsimulation modelling process

Table 8.1 The main demographic, socio-economic and health variables in SMILE

	Categorisation
Age	Continuous 0–100+
Sex	0 male; 1 female
Marital status	1 married; 2 separated; 3 divorced; 4 widowed; 5 never married
Household annual income (6-category dummy variable)	1 household income < €24,000; 2 household income < €30,000; 3 household income < €40,000; 4 household income < €50,000; 5 household income < €75,000; 6 household income > €75,001
Education level (7-category dummy variable)	1 no education; 2 primary and some secondary education; 3 junior certificate; 4 leaving certificate; 5 lower degree; 6 higher degree; 7 special needs education
Long-term illness (LTI)	0 no; 1 yes
GP utilisation	0 no; 1 yes
Medical card holder	0 no; 1 yes
Smoker or not	0 no; 1 yes
Owner of a private car or not	0 no; 1 yes

Spatial microsimulation is a method used to create spatially disaggregated microdata that previously did not exist. Thus, an important component of model development is validation as it is only through validation that the integrity of the model is established. SMILE contains a number of internal validation methods, namely, Z-scores and Z^2-scores (Hynes et al. 2009). The Z-score is based on the difference between the relative size of the category in the synthetic and actual populations, although an adjustment is made to the formula when dealing with zero counts. A Z-score can be summed and squared to provide a measure of tabular fit similar to a chi-squared statistic. If a cell's Z-score exceeds the critical value, the cell is deemed not to fit, while if a Z^2-score exceeds the critical value, then the dataset is deemed not to fit (i.e. $|Z| > 1.96$). The Z-score calculation is given by

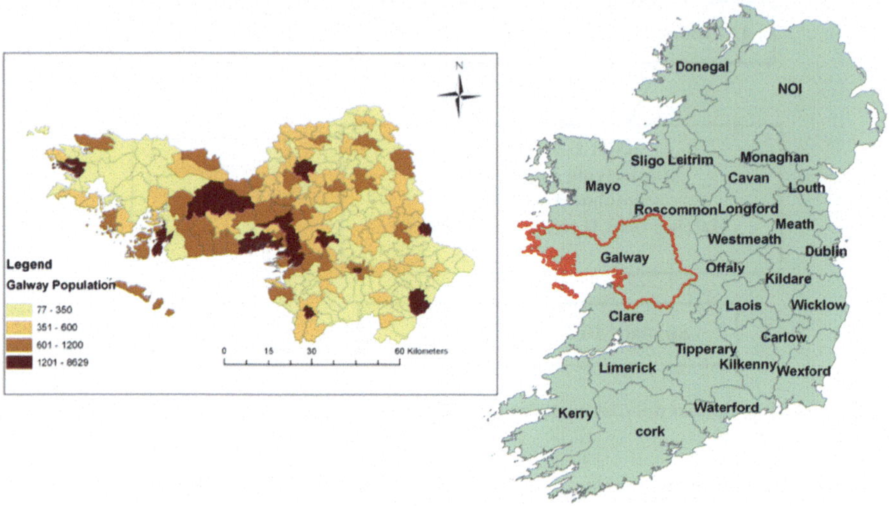

Fig. 8.2 Study area, Co. Galway

$$Z = \frac{\dfrac{T_{ij} - O_{ij}}{\sum_{ij} O_{ij}} \pm \dfrac{1}{2 \times \sum_{ij} O_{ij}}}{\sqrt{\dfrac{\left(\dfrac{O_{ij}}{\sum_{ij} O_{ij}}\right)\left(1 - \dfrac{O_{ij}}{\sum_{ij} O_{ij}}\right)}{\sum_{ij} O_{ij}}}}, \tag{8.1}$$

where T_{ij} is the estimated data, column *i,* row *j*, and O_{ij} is the census data. The $\frac{1}{2 \times \sum_{ij} O_{ij}}$ stochastic component is added or subtracted because in some large tables, it is possible to have 0 values, and then we would have division by zero. The stochastic component is added if $T_{ij} < O_{ij}$ and subtracted if $T_{ij} > O_{ij}$. If the observed and the expected values are the same, then *Z* is 0. The above formula is used to calculate the *Z*-score. It is easy to see from the sample of Z-squared results presented in Table 8.2 which tables and which EDs fit the best.

Information on the relative error and the *Z*-scores are outputted automatically in the static simulation. As shown in Table 8.2, the first line in section 3 of the table shows the associated 95% critical value for the Z^2-score. For illustration, the degrees of freedom are the number of columns in the table that represent education level. As there are four such columns, the associated degrees of freedom for specialist are 1.06. Taking ED 101004 as an example, the Z^2-score of zero indicates that the estimated tables fit the actual tables. Also for this ED, the *Z*-score is zero across all cells, indicating that the estimated cells fit the actual cells from the census perfectly.

Table 8.2 Comparison of simulated results to SAPS aggregates

Education	Education 1	Education 2	Education 3	Education 4	
1. Actual SAPS table					
101003	675	283	177	255	
101004	1,503	584	441	561	
101005	1,157	319	332	567	
101006	1,643	1,146	476	410	
2. Simulated table results					
101003	675	283	179	255	
101004	1,503	584	441	561	
101005	1,157	319	332	567	
101006	1,643	1,146	476	410	
3. Z-score	*X*-squared critical value, 1.06				Z^2-score
101003	0	0	0.16	0	0.002
101004	0	0	0	0	>0.001
101005	0	0	0	0	>0.001
101006	0	0	0	0	>0.001

On the other hand, in ED 101003, cell 3 is 0.16. This is above zero but still does not exceed the critical value, that is, these cells still fit the actual cells at the 95% confidence level, and its Z^2-score is also below the critical value (0.002), thus indicating that the estimated table still fits the actual table very well.

However, internal validation is only the first stage of the validation process. It is also necessary to compare the fitted model with 'fresh' external data. This external validation compares the newly created variables from SMILE to similar data that are not used in the original match (Caldwell and Keister 1996). The simulated LTI variable is externally validated against the special health module of the Quarterly National Household Survey (CSO, 2001). The QNHS is a representative dataset for the whole of Ireland. Producing variables at the NUTS-3 and county spatial levels (Table 8.3 provides a description of the population distribution of each spatial scale), it was found that the rates of LTI were similar between the SMILE dataset and the QNHS. It was found that SMILE predicted 26% of Galway's residents had a LTI, while the QNHS reported that 22% of Galway's population had a LTI. This indicated that the two datasets had similar rates for this indicator. Once validation of the synthetically created data is complete, the data produced by the model may be used for analysis with much more confidence.

This chapter links the output from SMILE to a SIM to examine whether ease of access to GP services has an impact on LTI in Ireland. The next section outlines the development of the spatial interaction model.

8.2.2 *Step 2: Service Provision – Accessibility Analyses Using Spatial Interaction Modelling*

To examine the effect access to GP services has on LTI, a spatial interaction model (SIM) was used to calculate 'access scores' (i.e. how easy, or otherwise, it is for a resident to access their GP) from each ED to the nearest GP. It is important to note

Table 8.3 The distribution details of the Irish population at the regional level, county level and ED level, 2002 (Source: CSO)

County population	Persons 2002	Percentage distribution (%)	County population	Persons 2002	Percentage distribution
Leitrim	25,799	0.7	Louth	101,821	2.6
Longford	31,068	0.8	Clare	103,277	2.6
Carlow	46,014	1.2	Wicklow	114,676	2.9
Monaghan	52,593	1.3	Wexford	116,596	3.0
Roscommon	53,774	1.4	Mayo	117,446	3.0
Cavan	56,546	1.4	Kerry	132,527	3.4
Sligo	58,200	1.5	Meath	134,005	3.4
Laoighis	58,774	1.5	Donegal	137,575	3.5
Tipperary North	61,010	1.6	Kildare	163,944	4.2
Offaly	63,663	1.6	Limerick City and County	175,304	4.5
Westmeath	71,858	1.8	Galway City and County	209,077	5.3
Tipperary South	79,121	2.0	Cork City and County	447,829	11.4
Waterford City and County	101,546	2.6	Dublin City and County	1,122,821	28.7
Kilkenny	80,339	2.1			
Regional population	*Persons 2002*	*Percentage distribution (%)*	*ED Irish average*	*ED Irish min*	*ED Irish max*
Border	432,534	11.0	889	55	14,238
Dublin	1,122,821	28.7	*ED Galway average*	*ED Galway min*	*ED Galway max*
Mid-east	412,625	10.5	719	77	8,629
Midland	225,363	5.8			
Mid-west	339,591	8.7			
South-east	423,616	10.8			
South-west	580,356	14.8			
West	380,297	9.7			

that there are no formal GP or primary care catchments in Ireland. Individuals may choose to visit whichever clinic they want. There are numerous methodologies available to measure accessibility to health-care services (Bertuglia et al. 1994 provide a comprehensive review of these methodologies). However, by far, the most popular methodology to be adopted has been the origin-constrained SIM (Clarke et al. 2002). A SIM may be built to describe and predict the flow of people, goods or services across space. They allow the modelling of the trade-off between spatial convenience and the attractiveness of particular destinations (measured by proxies such as size, brand and quality of the service). These models also allow the estimation of the number of trips to a particular destination, given the attributes or 'attractiveness' of that location. The attractiveness of an opportunity should be measured based on what characteristics of a potential destination are important to the consumer (Liu and Zhu 2004). For example, it can be measured as the number of retail outlets at a destination or the number of GPs in a single location or practice. In reality, the attractiveness of a facility often translates as its physical size (Birkin and Clarke 1991).

Of particular interest to this chapter, is that these models can be used to build a suite of performance indicators. Performance indicators allow the measurement of how well a particular service serves the residents (Clarke et al. 2002). Thus, these models quantify accessibility according to where individuals consume services as predicted by the SIM. SIMs, therefore, provide a more realistic representation of access to services than, for example, simply taking the number of service outlets in a zone or estimating accessibility through a simple straight-line nearest facility type indicator.

A SIM can be written as

$$T_{ij} = O_i A_i W_j \exp\left(-\beta d_{ij}\right), \tag{8.2}$$

where A_i is a balancing factor that ensures that

$$\sum_j T_{ij} = O_i. \tag{8.3}$$

A_i is calculated as

$$A_i = \frac{1}{\sum_j W_j \exp(-\beta d_{ij})}. \tag{8.4}$$

The residential zone *i* refers to the centroid of each ED, while the destination *j* refers to the *x*, *y* location of each GP centre. T_{ij} is the flow of individuals from each ED residential zone *i* to each GP centre *j*. The demand variable O_i is the number of residents with a LTI in each ED as simulated by SMILE. The attractiveness parameter for each health-care centre W_j, is the number of practitioners in each health centre (a measure of how easy it is to be examined quickly). The distance variable d_{ij} is the distance from each ED centroid (*i*) to each primary care service (*j*). d_{ij} was calculated using network analysis (using the current road data for Ireland) in ArcGIS.

As there is currently no data on interaction patterns to GP services in Ireland, a generic value for the distance-decay parameter was chosen. The distance-decay function value β is 0.2. Sensitivity analysis was conducted with the regard to the parameter value, using values ranging from 0.1 to 0.7. These results were not overly different, and consulting the literature on health service accessibility in the UK (Smith et al. 2006; Tomintz et al. 2008), β was assigned a value of 0.2. Taking the results from the SIM, two effectiveness indicators were then used to predict access scores for each ED (Clarke et al. 2002). The first effectiveness indicator calculates the aggregate level of provision for a particular origin zone i as follows:

$$w_i = \sum_j \frac{T_{ij}}{T_{*j}} W_j. \tag{8.5}$$

The above equation for estimating the aggregate level of provision for an area is calculated by dividing each SIM output Eq. 8.2 by the sum of all outputs for each zone j, where * indicates summation across all zones i. This is then multiplied by the attractiveness of zone j. The sum of all these values for residence zone i provides the aggregate provision for each zone i. This indicator ensures that even if an area does not have a service facility, the area will not have a zero accessibility score. Relating this aggregate provision indicator to the population in an area will allow the identification of areas where a significant number of households suffer poor accessibility to a particular service, in this instance, poor accessibility to a GP service centre. The level of provision per household is an indicator that divides the aggregate level of provision score, $w_{i,}$ by the number of households in the residence zone, i, as follows:

$$v_i = \frac{w_i}{I_i}. \tag{8.6}$$

Using both of these performance indicators prevents areas with a small population or no services being automatically labelled as 'poor access areas', when in fact these areas may reside close to neighbouring zones with good access (the interaction based indicators provide a type of smoothing effect). Figure 8.3 presents the spatial distribution of access scores to a GP for County Galway as calculated by the SIM and accessibility indicators. From Fig. 8.3, one can see that access is highest in the city and its hinterland, while access is lowest in the west and south-east of the county. Combining the results from the SIM within a logistic model (to examine the determinants of LTI) allows us to examine the impact of access to GP services on LTI at the sub-national level in Ireland.

8.3 Results: The Influences of Ill-Health at the Sub-national Level

Using the resulting dataset from these two models (i.e. SMILE and SIM), Table 8.4 provides an aggregate overview of three of the key individual-level drivers of LTI for County Galway. Table 8.4 indicates that on average, individuals with a LTI,

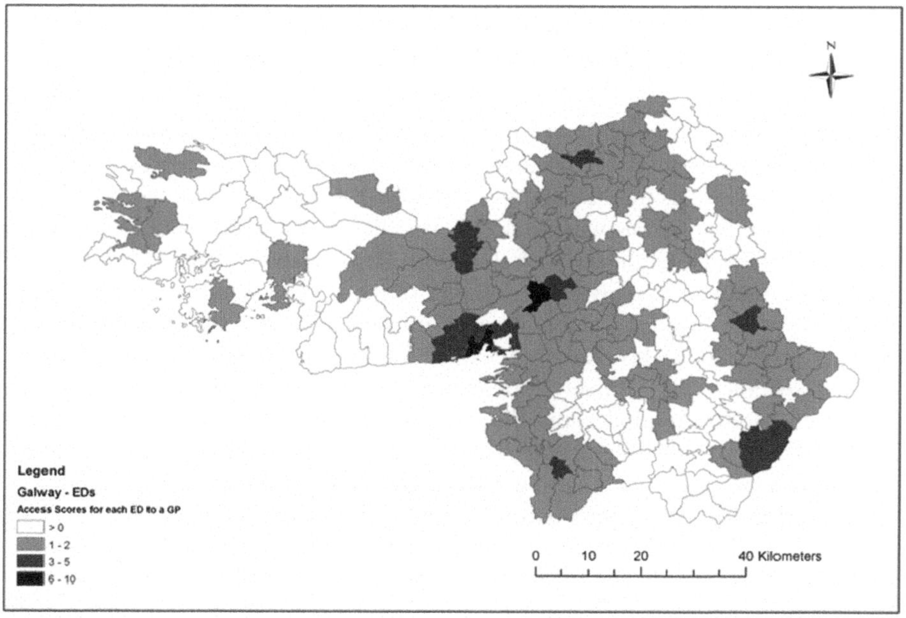

Fig. 8.3 Access scores for each ED to a GP service in County Galway as calculated by the accessibility indicators

Table 8.4 Descriptive statistics for three of the key individual-level drivers of LTI in County Galway

	No LTI reported	Reporting a LTI
Monthly household income	€19,833	€12,041
Age	38	50
Receive free medical care (%)	22	52
Access to GP services	3.50	2.50

compared to those not reporting a LTI, have lower income, are older, receive a higher percentage of free medical care and have poorer access to GP services.

In Ireland, an individual is entitled to a medical card (thus receiving free medical care) if their income is below a certain threshold, or they are over 70 years old, or they have a specific long-term illness. This was measured by whether the individual had a medical card (indicating they are in receipt of free medical care) or not. Thus medical card possession may be used as an indicator of deprivation in Ireland (Kelleher et al. 2002). It is well established in the literature that old age and lower income directly relate to poorer health status (Safaei 2007; Kelleher et al. 2002). Also included in Table 8.4 are the access scores to GP services for individuals with and without a LTI. A higher access score indicates increased ease of access for an individual, in terms of distance to the clinic and the number of GPs per clinic. Table 8.4 indicates that on average individuals without a LTI have higher access scores than individuals with a LTI, thus suffering from poorer access to GP services.

8.3.1 Step 3: Logistic Regression Modelling of the Relationship Between Ill-Health and Individual Characteristics

The relationship between ill-health (using LTI as a proxy for ill-health) and individual characteristics was examined by constructing a logistic regression model for County Galway using the resulting SMILE/SIM dataset. An initial set of uni-variate logistic regressions were undertaken to determine which variables had the greatest significant relationship with ill-health. The explanatory variables presented in Table 8.5 were chosen given the strength of their relationship with ill-health as found by previous national and international research. The variables included age (Sassi and Hurst 2008), sex (Kelleher et al. 2002), annual household income (dummy variable, see Table 8.1) (Safaei 2007), education (dummy variable, see Table 8.1) (Alborz et al. 2005), marital status (dummy variable, see Table 8.1) (Sassi and Hurst 2008), whether an individual smokes or not (Sassi and Hurst 2008), if the individual has visited a GP in the previous year (McGregor et al. 2008), and whether an individual is a medical card holder (entitled to free GP services) or not (Kelleher et al. 2002). The final variable to be included was whether the individual has access to a private car. This variable was included to examine the effect of accessibility on LTI. An average access score for each ED to a GP service was calculated by the suite of performance indicators outline above and matched into the SMILE dataset.

Table 8.5 presents the results (coefficient values, their associated standard errors, marginal effects and significance levels) of this multivariate logistic regression. On average, it is found that the main drivers of LTI in County Galway were as follows: special needs education (those that had special needs education requirements were 62% more likely to have a LTI), having visited a GP in the previous 12-month period (those that had attended a GP in the previous 12 months were 23% more likely to have a LTI), having an annual household income less than €24,000 (individuals with an income less than €24,000 were 11% more likely to have a LTI) and medical card/free medical care status (individuals with a medical card were 12% more likely to have a LTI).

Given the strong relationship between low income levels, medical card possession and LTI, this analysis is in line with other authors' work showing that income levels have a strong positive effect on health status in Galway (Safaei 2007). It is important to note that given the relationship between medical card possession and income, these two variables may be correlated and introduce issues of collinearity to the model. To check this, a correlation test for the two variables was run. The resulting correlation 0.24 (24%) was deemed low enough to include both variables in the analysis (a correlation of below 0.5 is generally deemed acceptable). With regard to having visited a GP in the proceeding 12-month period, previous research has indicated that individuals with LTI have higher rates of GP utilisation (McGregor et al. 2008). These results corroborate this, showing that visits to a GP are found to have a positive effect on the probability of an individual having a LTI. This study also found that individuals with a special needs education have a strong positive relationship with LTI. Physical and mental disabilities are often found to coexist for

Table 8.5 Micro-level determinants of ill-health, number of observations and chi-squared results

	Coefficient	Standard error	*P* value	Marginal effects
Education (dummy variable):				
No education (reference category higher degree)	0.8	0.05	<0.001	0.10
Primary and some secondary education (reference category higher degree)	0.66	0.05	<0.001	0.09
Junior certificate (reference category higher degree)	0.44	0.05	<0.001	0.06
Leaving certificate (reference category higher degree)	0.51	0.05	<0.001	0.73
Lower degree (reference category higher degree)	0.15	0.05	0.005	0.03
Special needs education (reference category higher degree)	3.27	0.17	<0.001	0.62
Household income (dummy variable):				
Household income < €24,0000 (reference category €15,000–23,999)	0.95	0.05	<0.001	0.11
Household income < €30,000 (reference category €15,000–23,999)	0.45	0.06	<0.001	0.70
Household income < €40,000 (reference category €15,000–23,999)	0.53	0.06	<0.001	0.09
Household income < €50,000 (reference category €15,000–23,999)	0.53	0.06	<0.001	0.09
Household income < €75,000 (reference category €15,000–23,999)	0.4	0.06	<0.001	0.07
Household income > €75,001 (reference category €15,000–23,999)	0.3	0.07	<0.001	0.04
Age	0.02	0.0003	<0.001	0.02
Sex	−0.4	0.01	<0.001	−0.06
Marital status (dummy variable):				
Marital status: separated (reference category married)	0.21	0.06	<0.001	0.02
Marital status: divorced (reference category married)	−0.09	0.14	0.53	−0.01
Marital status: widowed (reference category married)	0.09	0.04	0.02	0.007
Marital status: never married (reference category married)	−0.42	0.02	<0.001	−0.07
GP visit in previous 12 months	1.78	0.02	<0.001	0.23
Smoke	0.2	0.02	<0.001	0.02
Medical card status	0.68	0.01	<0.001	0.12
Own, or have access to, a car	−0.44	0.02	<0.001	−0.04
Access score	−0.02	0.002	<0.001	−0.02
Constant	−3.85	0.08	<0.001	
Number of observations (individuals aged 16 and over)	145,983			
Probability > chi^2	0			

individuals (Alborz et al. 2005). Research has found that individuals with learning disabilities do have higher GP utilisation rates (Alborz et al. 2005). It is interesting to note that while age was found to have a significant positive association with LTI, this effect was not large. Females and smokers were also found to have a higher probability of having a LTI than males and nonsmokers.

However, of particular interest to these analyses is the relationship between LTI and the access score, as calculated by the SIM outlined in Sect. 8.2.2. The access score is a continuous variable, where an increase in the score indicates an improvement in ease of access. Table 8.5 shows that the access score was found to have a significant, but small, negative effect on having a LTI (−0.02). That is, for every 0.02 decrease in ease of access to a GP service, an individual is more likely to have a LTI. These findings are consistent with Table 8.4, which shows that overall, individuals in Galway without a LTI have better access to services than individuals with a LTI. This would indicate that physical ease of access to health-care services can have a direct impact on an individual's health status. Another variable of interest that may be used as an indicator of ease of access to health services is private car ownership. Table 8.5 shows that private car ownership/availability has a significant negative relationship with LTI (individuals without a car are 4% more likely to have a LTI). Car owners are less likely to report suffering from a LTI than non-car owners. Further, Table 8.5 shows that although access to GP services was found to have a significant relationship with LTI, an individual's income level and deprivation level have a stronger effect on whether an individual has a LTI.

This analysis indicates that individuals with poorer access to GP services are more likely to report suffering from a LTI. However, without further time-series analysis, it is not possible to infer that poor access to GP services causes LTI. Previous research in the UK has found that increased distances to health services may be associated with poorer health service utilisation (Jordon et al. 2004, 2008, 2010). However, by endogenising the access score within the logistic model for LTI, this model establishes a clear relationship between access to health services and LTI. Finally, it is important to note that it is the combination of the SIM and the spatial microsimulation that allows this analysis to be carried out.

8.4 Discussion

Although the spatial disaggregation of both health service demand and supply is crucial in understanding health-care needs and service requirements, it is not always straightforward or possible within conventional health-care modelling frameworks. Difficulties arise primarily due to data limitations. In this chapter, we have attempted to improve the applicability of the data created by a microsimulation model, not by increasing the complexity of the simulation framework, but by integrating the newly created data within a meso-level, spatial interaction framework. Such an analysis provides policymakers with information on both demand for and supply of health-care services, thus allowing them to shape future service provision and target existing health resources in a more efficient and effective manner.

However, difficulties do arise when linking micro- and meso-level models. These issues include ensuring the consistency of the data being used between modelling frameworks and validation. First, in terms of data consistency, one must ensure that the data produced by the spatial microsimulation model can be aggregated to the required meso-level: that is, to ensure that both models have at least one common spatial scale. A second issue is model validation. As outlined above, the data provided by SMILE is validated across a range of parameters. Although trip data does not exist for individuals to GP services in Ireland and therefore cannot be validated, using best practice, an aggregate distance-decay parameter of 0.2 was used (Morrissey et al. 2010). Limitations aside, a major advantage of the modelling methodology, combining data from a spatial microsimulation model with the number of individuals with a LTI at the ED level within a spatial interaction model, is the production of a set of access scores for GP services for each individual in Co. Galway. These access scores were then used with other demographic and socio-economic variables to construct a logistic regression model of LTI to establish whether there was a relationship between an individual's access to GP services and their likelihood of having a LTI at the ED level in County Galway. Building on previous UK-based research that indicated that increases in distance to health-care facilities were associated with decreased levels of treatment for different illnesses, this chapter endogenises access to GP services as a determinant of LTI.

8.5 Conclusions

In conclusion, this chapter has found that inadequate service provision may be linked to LTI at the small area level in Ireland. These results are of interest to both the Irish and international policymakers, as they indicate that to ensure 'good' health provision across the general population, governments need to ensure better access levels to health services across space. Although the Department of Health and Children in Ireland has listed 'equity of access' to health services as one of its main objectives, this analysis has shown that access levels to GP services vary across space. Although disparities in health services in Ireland (and internationally) are primarily a function of the degree of urbanisation and the need to increase efficiency and effectiveness in the delivery of health-care services, particularly acute services (Asthana et al. 2002), the development of a framework to allocate health services in a more demand-driven manner is important.

The provision of the Irish health service is constantly under review, rationalisation and re-focus (McDaid et al. 2009). A framework that provides both demand and current supply-based analysis, as outlined above, allows policymakers to allocate scarce resources in a more effective manner. This is particularly relevant given the proposed roll-out of the primary care strategy and the need to allocate increased service provision to the primary care sector. As such, one may conclude that the Irish government needs to develop a framework for allocating health services on a more equitable basis given demand for these services across space. This chapter demonstrates how micro- and meso-level models may be combined to produce a more holistic analysis health-care demand and supply. Spatial microsimulation models, given their flexible scale disaggregation scale, provide a useful framework for linking meso-level models, such as SIM.

References

Alborz, A., McNally, R., & Glendinning, C. (2005). Access to healthcare for people with learning disabilities: Mapping the issues and reviewing the evidence. *Journal of Health Service Resource Policy, 10*(3), 173–182.

Asthana, S., Brigham, P., & Gibson, A. (2002). *Health resource allocation in England: What case can be made for rurality*. Plymouth: Rural Health Allocation Forum and the University of Plymouth.

Ballas, D., & Clarke, G. P. (2001). Modelling the local impacts of national social policies: A spatial microsimulation approach. *Environment and Planning C: Government and Policy, 19*(4), 587–606.

Ballas, D., Clarke, G. P., & Dewhurst, J. (2006). Modelling the socio-economic impacts of major job loss or gain at the local level: A spatial microsimulation framework. *Spatial Economic Analysis, 1*(1), 127–146.

Bertuglia, C. S. (1994). *Modelling the city: Performance, policy and planning*. Routledge: London.

Birkin, M., & Clarke, G. P. (1991). Spatial interaction in geography. *Geography Review, 4*, 16–24.

Caldwell, S., & Keister, L. (1996). Wealth in America: Family stock ownership and accumulation, 1960–1995. In G. P. Clarke (Ed.), *Microsimulation for urban and regional policy analysis* (pp. 88–116). London: Pion.

Campbell, N., Elliott, A., Sharp, L., Ritchie, L., Cassidy, J., & Little, J. (2000). Rural factors and survival from cancer: analysis of Scottish cancer registrations. *British Journal of Cancer, 82*, 1863–1866.

Clarke, G. P., Eyre, H., & Guy, C. (2002). Deriving indicators of access to food retail provision in British cities: Studies of Cardiff, Leeds and Bradford. *Urban Studies, 39*(11), 2041–2060.

Centrol Statistical Office. (2001). *Quarterly national health survey 2001*. Dublin: CSO.

Department of Health and Children. (2001). *Primary care a new direction*. Dublin: Department of Health and Children, Government of Ireland.

Dorling, D., Mitchell, R., & Pearce, J. (2007). The global impact of income inequality on health by age: An observational study. *British Medical Journal, 335*, 873–875.

Hynes, S., Morrissey, K., O'Donoghue, C., & Clarke, G. (2009). Building a static farm level spatial microsimulation model for rural development and agricultural policy analysis in Ireland. *International Journal of Environmental Technology and Management, 8*, 282–299.

Jones, A., Haynes, R., Sauerzapf, B., Crawford, M., Zhao, H., & Forman, D. (2008). Travel time to hospital and treatment for breast, colon, rectum, lung, ovary and prostate cancer. *European Journal of Cancer, 44*(9), 992–999.

Jones, A., Haynes, R., Sauerzapf, B., Crawford, M., & Forman, D. (2010). Geographical access to healthcare in Northern England and post mortem diagnosis of cancer. *Journal of Public Health, 32*(1), 1–6.

Jordon, H., Roderick, P., Martin, D., & Barnett, S. (2004). Distance, rurality and the need for care: Access to health services in SW England. *International Journal of Health Geographies, 3*, 21–30.

Kelleher, C., Harrington, J., & Friel, S. (2002). Measures of self-reported morbidity according to age, gender and general medical services eligibility in the national survey of lifestyle. *Attitudes and Nutrition, Irish Journal of Medical Science, 171*, 134–137.

Liu, S., & Zhu, X. (2004). An integrated GIS approach to accessibility analysis. *Transactions in GIS, 8*(1), 45–62.

Longley, P., & Batty, M. (2001). *Advanced spatial analysis: The CASA book of GIS*. Redlands: ESRI Press.

Lymer, S., Brown, L., Yap, M., & Harding, A. (2008). Regional disability estimates for New South Wales in 2001 using spatial microsimulation. *Applied Spatial Analysis and Policy, 1*(2), 99–116.

Lymer, S., Brown, L., Harding, A., & Yap, M. (2009). Predicting the need for aged care services at the small area level: the CAREMOD spatial microsimulation model. *International Journal of Microsimulation, 2*(2), 27–42.

Mackenbach, J. (2006). *Health inequalities: Europe in profile*. Brussels: European Commission. www.ec.europa.eu/health/ph_determinants/socio_economics/documents/ev_060302_rd06_en.pdf. Accessed 8 Nov 2008.

Madden, D. (2008). *Health and income poverty in Ireland, 2003–2006* (Health, Econometrics and Data Group (HEDG) Working Papers 08/14). HEDG, c/o Department of Economics, University of York.

McDaid, D., Wiley, M., Maresso, A., & Mossialos, E. (2009). Ireland: Health system review. *Health Systems in Transition, 11*(4), 1–268.

McGregor, P., McKee, P., & O'Neill, C. (2008). The role of non-need factors in individual GP utilisation analysis and their implications for the pursuance of equity: A cross-county comparison. *European Journal of Health Economics, 9*(1), 147–156.

Mitchell, R., Dorling, D., & Shaw, M. (2002). Population production and modelling mortality – An application of geographic information systems in health inequalities research. *Health and Place, 8*(1), 15–24.

Morrissey, K., Clarke, G., Ballas, D., Hynes, S., & O'Donoghue, C. (2008). Analysing access to GP services in rural Ireland using micro-level analysis. *Area, 40*(3), 354–364.

Morrissey, K., Hynes, S., Clarke, G., & O'Donoghue, C. (2010). Examining the factors associated with depression at the small area level in Ireland using spatial microsimulation techniques. *Irish Geography, 43*(1), 1–22.

Safaei, J. (2007). Income and health inequality across Canadian provinces. *Health and Place, 13*, 629–638.

Sassi, F., & Hurst, J. (2008). *The prevention of lifestyle-related chronic diseases: An economic framework* (OECD Health Working Paper 32). www.oecd.org/dataoecd/57/14/40324263.pdf. Accessed 12 Dec 2008.

Shaw, M., Smith, G., Thomas, B., & Dorling, D. (2008). *The Grim Reapers road Map: An atlas of mortality in Britain*. Bristol: The Policy Press, Health and Society Series: University of Bristol.

Smith, D., Clarke, G. P., Ransley, J., & Cade, J. (2006). Food access and health: A microsimulation framework for analysis. *Studies in Regional Science, 35*(4), 909–927.

Smith, D., Clarke, G. P., & Harland, K. (2009). Improving the synthetic data generation process in a spatial microsimulation model. *Environment and Planning A, 41*, 1251–1268.

Stillwell & Clarke, G. (2004). *Applied GIS and spatial analysis*. Chichtester: Wiley.

Tay, J., Kelleher, C., Hope, A., Barry, M., Nic, G. S., & Sixsmith, J. (2004). Influence of socio-demographic and neighbourhood factors on self rated health and quality of life in rural communities: findings from the Agriproject in the Republic of Ireland. *Journal of Epidemiology and Community Health, 58*, 904–911.

Tomintz, M., Clarke, G. P., & Rigby, J. (2008). The geography of schooling in Leeds: Estimating individual smoking rates and the implications for the location of stop smoking services. *Area, 40*(3), 341–353.

Vsc CSP Reference Quarterly Household National Survey. (2010). *Special module on heath*. Dublia: Central Statistics Office.

Wilkinson, R. G., & Pickett, K. E. (2006). Income inequality and population health: A review and explanation of the evidence. *Social Science and Medicine, 62*, 1768–1784.

World Health Organisation. (2000). *World health report 2000: Health systems: Improving performance*. Washington, DC: WHO.

Chapter 9
Projections Using a Static Spatial Microsimulation Model

Yogi Vidyattama and Robert Tanton

9.1 Introduction

So far, this book has described the process of spatial microsimulation modelling and has shown a number of different methods for creating static spatial microsimulation models. These static models provide estimates of small area statistics for the time period of the benchmark data – so if the benchmark data used is 2006 census data, then the reference period for the spatial model will be 2006. There is also a need to predict what an area will look like in the future, as a key issue in development planning is knowing where particular services will be required in the future. Therefore, a further development has been taken to project spatial microsimulation databases forward through time (see Harding et al. (2011) for a summary of key issues and spatial research using microsimulation).

Creating projections means attempting to 'age' the spatially microsimulated dataset. Therefore, a temporal dynamic has to be added to the model. As noted in Harding and Gupta (2007a), a conceptual distinction can be drawn here between spatial microsimulation models that undertake 'static ageing' (such as reweighting the small area dataset to future population projections) and those that attempt 'dynamic ageing', which involves updating the characteristics of the micro-units for each small area through time.

There are a number of spatial dynamic microsimulation models already in existence (e.g. SVERIGE and SMILE). There are also examples of pseudo-dynamic models in the UK, which are not fully dynamic in that they do not model individual life experiences like mortality, fertility and migration (as SVERIGE and SMILE do), but reweight to projections of census tables, so use static ageing. Examples of these models include SimBritain (Ballas et al. 2005a).

Y. Vidyattama (✉) • R. Tanton
National Centre for Social and Economic Modelling,
University of Canberra, Canberra, Australia
e-mail: Yogi.vidyattama@natsem.canberra.edu.au

R. Tanton and K.L. Edwards (eds.), *Spatial Microsimulation: A Reference Guide for Users*, Understanding Population Trends and Processes 6, DOI 10.1007/978-94-007-4623-7_9,

SVERIGE (Holm et al. 2001 and Chap. 12 of this book) uses the pattern of emigration, immigration, employment and earnings, education, leaving home, divorce, cohabitation and marriage, as well as mortality and fertility as dynamic individual behaviours in the model. A Monte Carlo simulation picks individuals in the microdata to experience any of the above behaviours based on simple probabilities and hence updates the individual characteristics in the microdata. This means that accurate probabilities of each behaviour are central to creating projections in this model. In SVERIGE, these probabilities are obtained using either probabilities from past experience or estimated logistic regression equations.

SMILE is built as both a static and dynamic spatial microsimulation model (Ballas et al. 2005b and Chap. 7 of this book). It is constructed to estimate and project small area statistics in Ireland. The model starts as a static model using an iterative proportional fitting (IPF) method to spatially disaggregate the aggregate microdata. Once this has been done, the demographic processes of mortality, fertility and migration are simulated. The mortality process is simulated by using the probability of death based on age, sex and location while the probability of birth is simulated based on age, marital status and location. The simulation of the migration process uses random sampling from calculated migration probabilities derived from the 1991 and 1996 census of population. These data provide migration probabilities from one area to another by age, sex and location.

SimBritain (Ballas et al. 2005a and Chap. 13 of this book) is a spatial microsimulation model for Britain's small areas. Unlike SVERIGE and SMILE, SimBritain is constructed as a pseudo-dynamic microsimulation model. The model projects benchmark tables from 2001 to 2011 and 2021 using the long-term trend of each small area based on data from the UK 1971, 1981 and 1991 census. The benchmark projections are calculated using a model of the changing population proportion in each category of each benchmark table. After all the six benchmark tables in SimBritain are projected, the microdata are reweighted to the projections, and new weights are calculated for each household or person on the microdata.

9.2 Projecting Small Areas Statistics in Australia Using a Spatial Microsimulation Model

SpatialMSM is a spatial microsimulation model that has been developed to estimate small area statistics in Australia and is described in Chap. 6. The model has been under development for several years, initially reweighting a household expenditure survey to 2001 census small area benchmarks (see Chin et al. 2005, 2006; Chin and Harding 2006, 2007 and, for documentation of the earliest models, see Melhuish et al. 2002).

As mentioned in Sect. 9.1, there are at least two ways to project an estimate using spatial microsimulation – static and dynamic ageing. At this stage, the development of the projections model in SpatialMSM has concentrated on static ageing, or a pseudo-dynamic process. There have been several approaches to static ageing that have been introduced in the model. First, a simple static ageing procedure has been adopted. This essentially involves inflating the weights for each household in the synthetic data using population projections for each small area and has been implemented in earlier work on projecting consumer characteristics out to 2020 in Australia (Harding and Gupta 2007b). This method can also be extended to using population projections by age and sex.

A second, more complex, approach is to project each one of the benchmark tables and then reweight to these new projections. This is conceptually similar to the approach followed in SimBritain (Ballas et al. 2005a). The approach used to project the benchmark tables leverages directly off small area population projections but can also use any other projections available. A third approach is to combine the first and second methods. This third approach can be used when there are not enough small areas available to implement the second method.

9.2.1 Projecting by Inflating the Microdata Weights

Because the survey data in SpatialMSM is benchmarked to the 2006 census (see Chap. 6), all the weights will refer to 2006 populations. One of the advantages of the reweighting method is that the weights can be easily inflated so that they will add up to a different year's population. This can be done using Statistical Local Area (SLA) level population projections from the Australian Bureau of Statistics (ABS) (ABS 2004). The formula used is

$$A_i^{sla}(2007) = A_i^{sla}(2006) * (Pop^{sla}_{2007} / Pop^{sla}_{2006})$$

where i is the record on the survey, sla is the SLA, $A_i^{sla}(2007)$ is the new weight, $A_i^{sla}(2006)$ is the weight benchmarked to 2006 census data, Pop_{2007}^{sla} is the population projection for that SLA in 2007 and Pop_{2006}^{sla} is the population of that SLA in 2006. This method assumes that the age/sex profile is constant over these years. For longer projections, where it would be reasonable to expect the age/sex profile to change, population projections by age/sex can be used to more accurately adjust the weights.

This method also assumes that any population growth is distributed to all the benchmarks at a constant rate over the years. If the population projections are by age and sex, then the assumption being made is that the relationship between the age/sex population projections and each of the benchmarks is constant over time.

This assumption that only the population size of each SLA is changing into the future can create problems. For instance, the long-term trend away from home ownership and towards private rental for younger generations will not be simulated because the population projections do not capture this change in preferences. Capturing this change would require data about such long-term trends at the small area level, and this is where the next method of projecting small area data can be used.

9.2.2 Projecting Each of the Benchmarks

One of the problems with the method outlined in Sect. 9.2.1 is that if weights are adjusted according to age/sex projections, then the inflated weights will be different for each person in a household. Typically, for weights on a survey, the weight for each person in a household is the same as the weight for the household. This is called 'integrated weighting'. It means that if all the weights for people in the household are summed, then the result will be the number of people in the household multiplied by the household weight. So inflating the weights using population projections by age and sex adjusts each person in the household differently, which means that the weights are no longer the same for each individual in the household, so the weights are no longer integrated.

The second method of projecting avoids this problem as the weights are recalculated using a process that incorporates integrated weighting. This method also incorporates projections of all the benchmark tables used in the reweighting process, providing a more accurate estimation process. This also means it is more flexible, as benchmark tables can be adjusted to incorporate different behavioural assumptions about the future.

This section describes how this method has been implemented in the SpatialMSM model, described in Chap. 6.

9.2.3 Projecting the Benchmark Tables

As described in Chap. 6, one of the first steps in the creation of SpatialMSM involves reweighting the two income survey sample files to benchmark tables from the 2006 census. Creating the outyear versions of the database again involves reweighting – but this time to newly created estimated benchmark tables for future years.

One of the advantages of reweighting to benchmark tables in future years is that the projected benchmark tables can be refined in the future based on new assumptions and knowledge that is available, and the weights can be easily recalculated using the refined benchmark tables. To get initial projections of each benchmark table, the model uses a log-linear regression model of each benchmark classification against age by sex by labour force status projections. One of the assumptions being made with this type of model is that people's behaviours and choices do not change in the future. If there is some knowledge about future behavioural changes, or if the user wants to model some future behavioural change to see the effect, then the benchmark tables could be refined and new weights calculated.

The benchmark tables used in this chapter are slightly different to those used in Chap. 6, as all the analysis in Chap. 6 was done with SpatialMSM/08B using 10 benchmarks, while the projections reported in this chapter were all done using a later model, SpatialMSM/08C, which split the benchmark for the number of people in a household into the number of adults in the household and the number of

Table 9.1 Benchmark tables used in order used

Benchmark table	Description	Type	Number of benchmark classes
1	Age by sex by labour force status	Person	32
2	Number of occupied private dwellings	Household	1
3	Dwelling tenure by weekly household rent	Household	6
4	Dwelling tenure by household type	Household	15
5	Dwelling structure by household family composition	Household	24
6	Household size – number of adults usually resident	Household	6
7	Household size – number of children usually resident	Household	5
8	Monthly household mortgage by weekly household income	Household	12
9	In different types of non-private dwelling	Person	4
10	Dwelling tenure by weekly household income	Household	25
11	Weekly household rental by weekly household income	Household	20
Total			150

Source: SpatialMSM/08C

children in the household. Apart from this slight change, the benchmarks are exactly the same as those used for Chap. 6. The list of benchmarks for SpatialMSM/08C is shown in Table 9.1.

The first benchmark table that is projected is the labour force by age by sex benchmark table, which has been projected up to 2027. To project this table, the SLA level population projections produced by the ABS for the Commonwealth Department of Health and Ageing (ABS-DOHA) (Department of Health and Ageing 2009) were combined with projections of labour force status used in the Australian Commonwealth Treasury's 2007 Intergenerational Report (IGR) (Treasury 2007). The long-run historical trend was also used in the report to project the participation rates for men and women of different ages. This means that the changing composition of the labour force in Australia is included in these projections, in particular the labour force changes in recent years with more women participating in the labour force.

Our initial problem with the ABS-DOHA SLA level population projections was that they are only available by age and sex and not by labour force status. Therefore, the projection of age by sex by labour force status was undertaken in two steps.

The first was to take the ABS-DOHA age by sex by SLA projections for 2007 (so the year after our benchmark tables 2006 reference year) and use the labour force by age/sex by SLA splits from the 2006 census data to apportion labour force status onto the 2007 age/sex population projections. The second step was then to use the percentage point change in the national projections of labour force status by age by sex from the Commonwealth Treasury's IGR 2007 report to adjust the proportion of persons in each labour force category for every SLA. It should be noted here that the national growth trend has been applied to each SLA, in the absence of any SLA-specific labour force projections.

The labour force by age by sex table then played an important role in the projections of all the other benchmarks since it was used as the exogenous variable that is then used to project the other benchmark tables. The projections for all the other benchmark tables were calculated using the relationship between the benchmark table being projected and the labour force by age by sex table in the base year (2006). The coefficients used to project all the other benchmark tables were estimated using a log-linear model:

$$Ln(PopBC) = f\left(\sum_{i=0\,j=1\,k=1}^{i=5\,j=6\,k=2} \beta_{ijk}\, Ln(PopLF_i\, Age_j\, Sx_k)\right) \tag{9.1}$$

where PopBC is the population in each benchmark table category while $PopLF_iAge_jSx_k$ is the population in labour force status i, age j and sex k. The estimation is done using a cross section regression with every SLA in Australia as an observation. Given that the estimate of β_{ijk} in Eq. 9.1 is the growth elasticity of the population in the benchmark table to the population in labour force status i, age j and sex k, the population growth in each benchmark table can be projected as

$$\frac{\Delta PopBC_{2006\text{-}T}}{PopBC_{2006}} = \sum_{i=0\,j=1\,k=1}^{i=5\,j=6\,k=2} \beta_{ijk} \frac{\Delta PopLF_i\, Age_j\, Sx_{k\,2006\text{-}T}}{PopLF_i\, Age_j\, Sx_{k\,2006}} \tag{9.2}$$

The estimation in Eq. 9.2 will give the estimated growth and hence the estimated number of every category's population in the benchmark tables for any year into the future. Note that all the financial data has been kept in 2006 prices, so we have not inflated rents, mortgages, incomes, etc. What we are projecting is the number of people in each income category or the number of people in each rent category. So the categories stay the same each year; only the number of people in each category changes.

To derive reasonable estimates from Eq. 9.2, the total number of people or households in each benchmark table must be the same. In many cases (due to randomisation by the ABS), these totals are not the same. Therefore, the number of people or households in each table is adjusted, so the totals are the same across all benchmark tables. This adjustment process takes one table as having the correct number and then adjusts all the other tables so they match this first table. In this case, the tables used as the basis on which to match the totals in all other tables were benchmark table number 1 for persons and benchmark table number 2 for households (Table 9.1). This means that there is an assumption that benchmark table number 1 has the

Table 9.2 R^2 for benchmarks used in the reweighting algorithm

Table no.	Benchmark table	Lowest R^2	Highest R^2	Mean R^2
2	Total number of households by dwelling type (occupied private dwelling/non-private dwelling)	0.542	0.993	0.767
3	Tenure by weekly household rent	0.424	0.862	0.635
4	Tenure by household type	0.516	0.984	0.826
5	Dwelling structure by household family composition	0.386	0.975	0.706
6	Number of adults usually resident in household	0.952	0.995	0.971
7	Number of kids usually resident in household	0.957	0.997	0.977
8	Monthly household mortgage by weekly household income	0.176	0.928	0.643
9	Persons in non-private dwelling	0.295	0.719	0.420
10	Tenure type by weekly household income	0.428	0.977	0.760
11	Weekly household rent by weekly household income	0.136	0.825	0.598

Source: Authors' calculations, SpatialMSM/08C

correct total for number of people, and benchmark table number 2 has the correct total for total number of households. All other tables are then adjusted to match the totals in these tables.

9.2.3.1 Reliability of the Projected Benchmarks

After the weights for future years are produced, the next step is to check the reliability of the estimation using this set of future weights.

There are two sources of model error in our projections. One comes from the projections of each benchmark table, so it has something to do with the reliability of the coefficient β_{ijk} in Eq. 9.1. The second source of error is in the generalised regression routine that reweights the survey data to the projected benchmarks.

In terms of the first source of model error, if the age by sex by labour force projection is not very good at estimating our other benchmarks, then the estimated weights for the projections will be inaccurate and the projections will be unreliable.

The estimate of the size of the errors in the forecasting of the benchmarks can be looked at using the coefficient of determination (R^2) of the regression process that produces the elasticity coefficients in Eq. 9.1. This figure will show how much variation in the benchmark table in the base year can be explained by the age by sex by labour force structure. As the regression was done for each category in each benchmark table, each of these will have its own R^2. To simplify the analysis, the means of the R^2 in the benchmark tables have been presented. The range of R^2 values is also provided to give a better idea as to the reliability. The results are shown in Table 9.2.

Table 9.3 Number of SLAs dropped due to failed accuracy criteria in SpatialMSM/08C

State/ territory	SLAs with failed accuracy criteria	Total SLAs	Percentage of SLAs with failed accuracy criteria (%)	Percentage of all persons living in SLAs with failed accuracy criteria (%)
NSW	2	200	1.0	0.4
VIC	4	210	1.9	0.0
QLD	43	479	9.0	0.8
SA	7	128	5.5	0.4
WA	17	156	10.9	0.9
TAS	1	44	2.3	0.1
NT	48	96	50.0	25.2
ACT	16	109	14.7	1.0
Australia	138	1,422	9.7	0.7

Looking at Table 9.2, the R^2 indicates that most of the variation in the original tables can be explained by the age by sex by labour force status table. This means that projections of these benchmarks tables using a coefficient calculated in the base year, while not perfect, would be reasonable as a first attempt at projecting the base microdata. Further work could enhance these projections, and one option may be to introduce some historical time series where the projections are particularly bad (as has been done for SimBritain; see Ballas et al. 2005a), but for most of the benchmarks, the age by sex by labour force status table explained on average more than 70% of the variation in the other tables. However, there are three tables where the average R^2 was below 70%, which are tenure by weekly household rent, monthly household mortgage by weekly household income and weekly household rent by weekly household income. This means that further work could be conducted on getting better projections in terms of housing cost and income.

In conclusion, on the basis of the R^2 for the model in Eq. 9.1, it is considered that the projected benchmarks were reliable enough to use in the reweighting process.

9.2.3.2 Reweighting to the Projected Benchmark Tables

The reweighting process is the same as that described in Chap. 6 but with different benchmark tables (as discussed above). One of the problems with using this technique is the loss of some SLAs because of failed accuracy criteria, so the procedure failed to find a solution given the constraints from the 11 benchmark tables and a limit on the number of iterations. The SpatialMSM model used for the base year (2006) produced weights for 1,214 SLAs and failed to produce reliable weights (so the accuracy criteria failed – see Chap. 6) for 138 SLAs. Most of the areas where the accuracy criteria failed were industrial areas, office areas or military bases with very low population counts. As a result, the proportion of people living in these SLAs is very small (Table 9.3). Only 0.7% of the total Australian population in 2006 were lost in the reweighting process.

The results from the reweighting process for the projected benchmarks shows that the further the model is projecting out, the more SLAs fail the accuracy criteria. In the base year of SpatialMSM/08C, there are 138 out of 1,422 SLAs in the base year failed the accuracy criteria. The number of SLAs that failed the accuracy criteria increases to 157 out of 1,415 SLAs in the 2010 projection and increases further to 208 SLAs and 236 SLAs in the 2020 and 2027 projections, respectively. Table 9.4 shows that besides the Australian Capital Territory and Northern Territory, most of the additional SLAs that failed the accuracy criteria are non-capital city SLAs.

Losing 236 of the 1,415 SLAs in the 2027 projection is still considered as acceptable for the purposes of this study since these SLAs only contain 2.8% of the whole population (Table 9.5). It should be noted, however, that around one-quarter to one-third of the Australian Capital Territory and Northern Territory populations live in SLAs which failed our accuracy criteria test in 2027, so projections for the two territories must be treated with caution.

One of the methods of validation for spatial microsimulation models described in Chap. 15 checks the accuracy of the estimated results from a spatial microsimulation model against a variable that is not benchmarked, but is available from some small area data source and that is accurate for the small areas being estimated. This method can also be applied to a model that projects data, as long as projections of small area data are available. In our case, the number of children aged 3 and 4 years is not benchmarked (we benchmark the number of children aged 0–17 years), can be estimated from our model and is available from the age/sex projections.

The measure described in Chap. 15 to assess this accuracy is the Standard Error around Identity (SEI). The SEI for SpatialMSM/08C in the base year (2006) is 99.0% for number of children aged 3–4, so we get an excellent result for the base year. The SEI for the projection in 2027 is 95.1%. This shows that the projected data match very well to the ABS population projections.

9.2.4 Projecting the Benchmarks When There Are Only a Small Number of Areas

The main advantage of a projection method based on projecting the benchmark tables compared to inflating the weights is that the former projection methodology allows the user to decide how the benchmark tables are projected forward. In the example used here, we have used a regression model that projects forward all the benchmark tables using age, sex and labour force status, but more complex methods could be used for different tables. Inflating the weights does not re-benchmark the future weights to any projected benchmark tables, so any growth in the area is purely based on population growth, by age and sex if required.

However, there are problems with this method, particularly when only a small number of areas are being estimated in the model. The reason is that the regression model used to project the benchmarks has each area as one observation for the regression. The results for the regression will therefore be unreliable if there are not many areas to estimate the model with.

One way to solve this problem is to combine the approaches of inflating the weights and projecting the benchmarks. So the benchmark tables can be projected by inflating the weights from the SpatialMSM model in the base year (i.e. 2006).

Table 9.4 Number of SLAs dropped due to failed accuracy criteria in the projections by major statistical region

Major statistical region (MSR)	SLAs which failed the accuracy criteria in SpatialMSM/08c	Total SLAs projected	SLAs with failed accuracy criteria in 2010 projection	SLAs with failed accuracy criteria in 2020 projection	SLAs with failed accuracy criteria in 2027 projection
Sydney	1	64	0	0	0
NSW-Balance of State	1	135	2	10	15
Melbourne	0	79	2	2	2
VIC-Balance of State	4	130	7	14	25
Brisbane	3	215	7	6	8
QLD-Balance of State	40	263	40	48	46
Adelaide	0	55	0	0	0
SA-Balance of State	7	72	10	18	20
Perth	2	37	2	1	2
WA-Balance of State	15	118	17	24	27
Hobart	0	8	1	1	1
TAS-Balance of State	1	35	2	3	3
Darwin	6	41	6	10	12
NT-Balance of State	42	54	43	44	43
Canberra	15	108	17	26	31
ACT-Balance of State	1	1	1	1	1
Australia	138	1,415	157	208	236

Source: SpatialMSM/08C projections

Table 9.5 Number of SLAs dropped due to failed accuracy criteria in the 2027 projection

State/ territory	SLAs with failed accuracy criteria	Total SLAs	Percentage of SLAs with failed accuracy criteria (%)	Percentage of all persons living in SLAs with failed accuracy criteria (%)
NSW	15	199	7.5	1.6
VIC	27	209	12.9	2.6
QLD	54	478	11.3	2.3
SA	20	127	15.7	3.4
WA	29	155	18.7	1.6
TAS	4	43	9.3	2.5
NT	55	95	57.9	32.5
ACT	32	109	29.4	24.7
Australia	236	1,415	16.7	2.8

Source: SpatialMSM/08C projections

For example, we have the growth in the population by age and sex in each SLA from population projections produced by the ABS for DOHA (Department of Health and Ageing 2009), and these can be used as inflation factors for every individual in the age group and sex. However, this will now mean that the household weight will no longer equal the person level weights, which is always the case when integrated weighting has been used, but we can then aggregate the individual weights to produce benchmark tables at the person level for each SLA and take the average of every individual in one household and aggregate to SLA level to produce the benchmark table at a household level. These tables are then used in the reweighting process. It is also necessary to ensure that the total number of people or households in each benchmark table is the same using the method outlined in Sect. 9.2.3.

The main advantage of this approach is the simplicity of the benchmark projection, which uses the results of the SpatialMSM model in the base year to project the benchmark tables. However, the problem with it is that we have no idea of whether the relationship between the SpatialMSM model in the base year and the projected benchmarks would hold in the longer term, as it is based on only one observation for a specific area.

9.2.5 Uprating Income and Cost

The three projection approaches described above assume that all prices including wages and salary in this model stay constant. Therefore, up to this point, the model is designed to project the composition of the population based on dwelling tenure type or family type rather than projecting financial variables such as poverty or housing stress. So the final part of the projections model is to project increases in the financial information.

In Australia, there are several databases from the ABS that can be used to inflate or uprate financial data into the future. Ideally, the uprating should be done using projections of the inflation factors provided by the government or other sources; however, these are rarely available for a number of years into the future, so the best option may be to use some trend over the last 10 years.

In the case of the projections from SpatialMSM, income is inflated using the average weekly earnings (AWE) data up to the end period available from the government (2010). Longer periods are predicted using the 10-year trend in the AWE. The AWE is based on a survey of payments paid by various businesses to their employees. Because the data are based on an ABS survey and the survey is a business survey, not a household survey, the ABS cannot produce estimates for areas below state. The projections are calculated for each state and territory using the trend from 10 years of data, and these are then used to calculate income projections for each SLA in that state or territory.

To make the estimates better, it is also possible to specify different rates for different genders, different industries of employment and also for full-time and part-time workers.

Another data source that can be used to inflate incomes is the Survey of Income and Housing, which can be used to get a different rate of increase for different income quintiles. This captures any changes in the income distribution, which we know changes over time (Vu et al. 2008).

To inflate costs, the Australian consumer price index data from the ABS provides a rate of increase in general costs. One of the main costs used in estimating housing stress in SpatialMSM is housing costs. The two housing costs used in the model are the housing costs of those who are renting and of those who are paying a mortgage. While the rate of increase for the former can be calculated directly from the rent component of CPI that is available as an expenditure class component in the CPI publication from the ABS, the increase in mortgages is more complicated. This is because there are at least three factors determining the mortgage payment – the house price, the length of mortgage and the interest rate. In many statistical software packages, there is a MORT function that can calculate a mortgage payment given an interest rate, the house price and the length of the mortgage.

If we assume the length of the mortgage is constant for all mortgagees as 30 years, we can then estimate the change in the mortgage payment given any projected future change in interest rates and house prices. This is the method used to project mortgage costs in the SpatialMSM model.

Note that for all these inflation factors, state figures have been used to inflate small area values. This is because there is no information on small area inflation factors in Australia.

9.3 Results

Results from the projection method described in Sect. 9.2 are excellent. In a paper describing the method and validation of the method (Vidyattama and Tanton 2010), which created projections for a number of variables out to 2027, the results showed a very good level of correlation with official population projections for small areas. The Standard Error around Identity (SEI) for the 2027 projections, using the projections of the number of children aged 3–4 from the model and from the DOHA projections (Department of Health and Ageing 2009), was 95.1%. This is a very high level of correlation, showing excellent results from the model (for further results see Harding et al. (2011)).

One of the main advantages of the spatial microsimulation projection method is that cross tabulations for the projections can now be derived. These can be very useful for planners looking at where services will be required in the future. As an example, Vidyattama and Tanton (2010) show in each small area the proportion of children aged 3–4 with both parents working, giving insights into where childcare places will be required in the future.

9.4 Strengths and Weaknesses

In this section, we sum up the strengths and weaknesses of the various projection methods outlined above.

The strength of the first method, inflating the weights using age/sex population projections, is that it is simple. Having said this, this simplicity is also its weakness, in that it does not provide the accuracy or the flexibility that the second method provides. While it is reasonable as a first attempt, it fails to take into account anything except changes in the age/sex population. It also leads to different weights for people and households, which means it is not possible to sum the person level weights in a household, divide by the number of people in the household and get the household weight.

The main strength of the second method (projecting the benchmark tables) is the ability to provide a picture of a household's future characteristics according to assumptions given by other models, such as population projections from the ABS and labour force projections from the Intergenerational Report. While age, sex and labour force projections are the main determinants of the projections from the current SpatialMSM model, there is the capacity to include other projections as they become available and incorporate these quickly into the modelling by benchmarking to the revised projected tables. It is important to note that the performance of this model is highly dependent on the assumptions and the performance of the projections being used to model the benchmark tables.

Another strength of this projection method is the independence of each SLA in the model. This means that each SLA can have a scenario change applied separately, and as long as the SLA does not fail the accuracy criteria (so we are getting reasonable estimates in the future), then the model can provide projections for that SLA. However, this feature is also one of the weaknesses of this projection method, as those SLAs may interact through population movement, especially if unemployment rates are changed in one SLA, and this population movement is not modelled (although it could be in the future through a dynamic model).

The main weakness of the method is the fact that the projection relies on the relationship between the labour force by age by sex composition and the composition of the other benchmark tables based on the 2006 population census. This is a reasonable assumption if the model projects into the near future, but unlikely for any long-term projection. Any change in personal preferences could make this assumption invalid. For example, one change in the housing industry that we may expect to see is people preferring to rent instead of buying their own house in the future, due to labour mobility and the increasing cost of purchasing a house. As a result, even if there are no changes in the structure of the labour force by age by sex, the proportion of younger people who live in rental accommodation may be increasing in the future, and this will not be captured in this model.

Not only are the benchmarks based on 2006 census data projected forward, the survey data used in this version of SpatialMSM is from 2002/2003 to 2003/2004. There is a strong possibility that these data do not represent individual households in the long term. Cassells and Harding (2007) show that the generation born between 1976 and 1991 (generation Y) has different characteristic to the previous generation in terms of working and having families, which are two variables we benchmark to. Because this is a static microsimulation model, we are not ageing the population at all; we are just benchmarking this 2002/2003 and 2003/2004 data to future projections. So in 2002, Gen Y is aged between 11 and 26. The characteristics in the benchmarks for this age group are projected into the future, and then, people aged 11–26 in 2027 will be benchmarked to these tables. So we are applying the Gen Y characteristics to people aged 11–26 in 2027. But people aged 11–26 in 2027 may be very different from the Gen Y group in 2002. Further, the Gen Y group from 2002 will be aged between 36 and 51 in 2027, and their characteristics may be very different from people aged between 36 and 51 in 2002.

Again, the flexibility of this model means we could assume some other preferences for this group in 2027 and adjust the benchmark tables using some behavioural model, but we really have no information on what preferences these people will have in 2027. So using the preferences from 2006 may be the best information available.

The advantage of the third method is that it can be used with a small number of areas. Using the second method, there needs to be a number of areas to run a regression of age by sex by labour force status against all the other benchmarks. This is not required for the third method. This also means that the projections are based on a small number of homogeneous areas, rather than a large number of heterogeneous areas, so the projections should be better.

The disadvantage of the third method is that there is a limited sample for projecting the benchmarks, and so there is not much to validate against.

9.5 Conclusions

This chapter has given an overview of a model that can address the need for small area information not only for the present but also for the future. In the past decade, this need has become more and more apparent as planning agencies in Australia (such as its local and federal governments) need to focus on service delivery for local areas given the characteristics of individuals and households in those areas (see Harding et al. 2011; Lymer et al. 2008, 2009).

A static ageing process is the approach taken in developing the projection model given the very high degree of complexity, cost and data requirements to build a fully dynamic microsimulation model. The static ageing model is undertaken by employing the currently available population and labour force projections to estimate the various constraint tables used in SpatialMSM/08C. The model then used the reweighting process in SpatialMSM/08C to allocate the microdata or unit record data according to the projected constraints.

This method has been able to produce information for small area planning into the future with a reasonable degree of reliability. The model is also able to take some simple scenarios to model some changes in the future and seems to be most reliable for capital cities. Nevertheless, the static ageing approach that the model uses means that it is difficult to model any behavioural change, without identifying the effect of the behavioural change and implementing this in the benchmark tables. Further, while we have not tested this, we expect that any large changes in the characteristics of the society in the future will be difficult to estimate, as the large changes in the benchmarks will mean the reweighting process will fail to find reasonable weights for a high proportion of areas.

References

ABS. (2004). Population projections for all States and Territories, Statistical Local Area, 2002 to 2022, Special Request Table from ABS Demography Area.

Ballas, D., Clarke, G., Dorling, D., Eyre, H., Thomas, B., & Rossitor, D. (2005a). Simbritain: A spatial microsimulation approach to population dynamics. *Population, Space & Place, 11*(1), 13–34.

Ballas, D., Clarke, G., & Weimers, E. (2005b). Building a dynamic spatial microsimulation model for Ireland. *Population, Space and Place, 11*, 157–172.

Cassells, R., & Harding, A. (2007). *Generation whY?* (AMP NATSEM Income and Wealth Report). Issue 17, AMP, Sydney.

Chin, S. F., & Harding, A. (2006). *Regional dimensions: Creating synthetic small-area microdata and spatial microsimulation models* (NATSEM Technical Paper no. 33). University of Canberra, Canberra. https://guard.canberra.edu.au/natsem/index.php?mode=download&file_id=648. Accessed 18 Sept 2011.

Chin, S. F., & Harding, A. (2007). SpatialMSM – NATSEM's small area household model of Australia. In A. Harding and A. Gupta (Eds.), *Modelling our future: Population ageing, health and aged care, International Symposia in Economic Theory and Econometrics*. Amsterdam: North Holland.

Chin, S. F., Harding, A., Lloyd, R., McNamara, J., Phillips, B., & Vu, Q. N. (2005). Spatial microsimulation using synthetic small-area estimates of income, tax and social security benefits. *Australasian Journal of Regional Studies, 11*(3), 303–336.

Chin, S. F., Harding, A., & Bill, A. (2006). *Regional dimensions: Preparation of the 1998–99 household expenditure survey for reweighting to small-area benchmarks* (Technical Paper no. 34). NATSEM, University of Canberra, Canberra. https://guard.canberra.edu.au/natsem/index.php?mode=download&file_id=649. Accessed 18 Sept 2011.

Department of Health and Ageing (2009). *The Australian population, statistical local area population projections, 2007 to 2027*, Revised. http://www.health.gov.au/internet/main/publishing.nsf/Content/ageing-stats-lapp.htm. Accessed 25 Jan 2011.

Harding, A., & Gupta, A. (2007a). Introduction and overview. In A. Harding and A. Gupta (Eds.), *Modelling our future: Population ageing, social security and taxation, International Symposia in Economic Theory and Econometrics*. Amsterdam: North Holland.

Harding, A., & Gupta, A. (Eds.). (2007b). *Modelling our future: Population ageing, social security and taxation. International Symposia in Economic Theory and Econometrics*. Amsterdam: North Holland.

Harding, A., Vidyattama, Y., & Tanton, R. (2011). Demographic change and the needs-based planning of government services: Projecting small area populations using spatial microsimulation. *The Journal of Population Research, 28*(2–3), 203–224.

Holm, E., Holme, K., Makila, K., Kauppi, M. M., & Mortvik, G. (2001). *The SVERIGE spatial microsimulation model – Content, validation, and example applications*. Kiruna: Spatial Modelling Centre, Umeå University.

Lymer, S., Brown, L., Yap, M., & Harding, A. (2008). Regional disability estimates for New South Wales in 2001 using spatial microsimulation. *Applied Spatial Analysis and Policy, 1*(2), 99–116.

Lymer, S., Brown, L., Harding, A., & Yap, M. (2009). Predicting the need for aged care services at the small area level: The CAREMOD spatial microsimulation model. *International Journal of Microsimulation, 2*(2), 27–42.

Melhuish, T., Blake, M., & Day, S. (2002, 29 September–2 October). *An evaluation of synthetic household populations for census collection districts created using spatial microsimulation techniques*. 26th Australian and New Zealand Regional Science Association International (ANZRSAI) Annual Conference, Gold Coast, Queensland, Australia.

Treasury (2007). *Intergenerational report 2007*. Australian Commonwealth Department of Treasury, Canberra. http://www.treasury.gov.au/igr. Accessed 30 Aug 2011.

Vidyattama, Y., & Tanton, R. (2010). Projecting small area statistics with Australian spatial microsimulation model (SpatialMSM). *Australian Journal of Regional Studies, 16*(1), 99–126.

Vu, Q. N., Harding, A., Tanton, R., Nepal, B., & Vidyattama, Y. (2008). *AMP.NATSEM income and wealth report 20 – Advance Australia fair?* Sydney: AMP.

Chapter 10
Limits of Static Spatial Microsimulation Models

Robert Tanton and Kimberley L. Edwards

This chapter outlines some of the limits of static spatial microsimulation models, covering data limitations; how adding different benchmark tables affects the results; issues with non-converging areas when too many benchmark tables are specified; and how a non-representative sample for the survey data affects results.

10.1 The Limitations Considered

There are a number of limitations of spatial microsimulation models discussed in detail in this chapter. Note that these are not all the limitations of spatial microsimulation models; these are just some that research has been reported on (see Tanton and Vidyattama 2010). These limitations are:

1. Data limitations
2. The effects of adding benchmark tables
3. The representativeness of the survey data for smaller capital cities

10.1.1 Data Limitations

One of the main limitations of static spatial microsimulation models is around the data requirements. For a spatial microsimulation model, three things are required:

R. Tanton (✉)
National Centre for Social and Economic Modelling, University of Canberra, Canberra, Australia
e-mail: Robert.tanton@natsem.canberra.edu.au

K.L. Edwards
School of Clinical Sciences, University of Nottingham, Nottingham, UK
e-mail: Kimberley.edwards@nottingham.ac.uk

R. Tanton and K.L. Edwards (eds.), *Spatial Microsimulation: A Reference Guide for Users*, Understanding Population Trends and Processes 6, DOI 10.1007/978-94-007-4623-7_10,

1. Record unit survey data to benchmark
2. Reliable small area data to benchmark the survey data
3. The same definitions for each variable on each of the two datasets

As an example, with the Australian SpatialMSM model described in Chap. 6, problems were experienced in the early developmental stages of the model because the reliable small area data from the census included people in non-classifiable households, whereas the survey data did not. To correct for this, special benchmark tables were requested from the Australian Bureau of Statistics which excluded non-classifiable households from the census benchmark data.

Non-matching variables between the census and survey may also mean that some variables cannot be benchmarked due to definitional differences. As an example, in Australia, some earlier surveys and the census before 2001 collected data on age left school (e.g. 15, 16 or 17), while current data collections collect data on highest year of school completed (year 11, year 12, etc.). This means that this variable could not be benchmarked if these datasets were used.

In other cases, some aggregation may be required to be able to benchmark the variable, so for example, on the Australian census, the landlord type has different classifications to those in the 2007–2008 Survey of Income and Housing. The census has eight valid landlord-type categories (excluding not stated and not applicable), whereas the 2007–2008 Survey of Income and Housing has six valid categories (excluding not applicable). The census includes categories on employers, housing co-operatives, etc., which the survey does not include – all these are included in an 'Other' category. Therefore, to make them comparable, these categories in the census need to be summed into one 'Other' category which then can be benchmarked to the survey 'Other' category.

In some ways, this can be overcome by using synthetic data as the basis for the model rather than actual survey data. This synthetic data can be created using small area probabilities, but the final model is not as accurate as it would have been if real survey data had been used. An example of model that has used synthetic unit record datasets is SYNTHESIS (Birkin and Clarke 1988, 1989).

10.1.2 The Effects of Adding Benchmark Tables

For all static spatial microsimulation models, as more constraint tables are added, the procedure doing the estimation will have greater difficulty matching all the benchmark tables. This is because the complexity of the algorithm being used (whether IPF, CO or generalised regression reweighting) has increased – so for example, for the generalised regression spatial microsimulation model described in Chap. 6, the constraints (X^c) are met using a constrained optimisation function. As the number of constraint tables increases, this constrained optimisation function gets more complicated and has a higher likelihood of not converging.

The question then is why not use an absolute minimum number of benchmarks. The trouble with this is that a smaller number of benchmarks mean there is less to constrain to, and so the final weights may not provide the accuracy required. As a test of reducing the number of benchmarks, a regression analysis was conducted of poverty rates against a set of ten benchmark variables. This regression showed that only six benchmark variables were significantly correlated with poverty at the 5% level, so eight benchmark tables were constructed from these six benchmark variables, and the SpatialMSM model described in Chap. 6 was run with these eight benchmark tables. The results from this version of the SpatialMSM model were that the number of areas passing the accuracy criteria increased from 1,302 to 1,337, so there were an additional 35 areas with an acceptable TAE (see Chap. 6). However, the accuracy of the model dropped from a standard error around identity (SEI – see Chap. 15) of 0.94 to an SEI of 0.82. This means that removing two benchmark tables led to a reduction in the accuracy of the model. So reducing the number of variables led to more areas technically being passed in the model, but a much lower accuracy.

The other advantage of increasing the number of benchmark tables is that the model becomes more generalisable. For example, including education in the benchmark variables means that cross tabulations with education in them can be derived, so poverty for people who's highest level of education was school compared to having a university degree.

In Tanton and Vidyattama (2010), a number of benchmark tables were added to the SpatialMSM model, and the results were tested. This analysis showed that from a base model with 11 benchmarks, 1,284 areas which passed the accuracy criteria and an SEI of 0.93, adding two additional benchmarks (education and occupation) led to a reduction in the number of areas that passed the accuracy criteria to 1,257 and an increase in the SEI to 0.94. So an additional 27 SLAs were unusable, but the accuracy of the model increased, and it was more generalisable (so poverty rates for different occupations could be calculated).

The appropriate number of benchmarks to be used also depends on the strength of the relationship between the benchmark variable(s) and the output variable(s). If extra tables are added where these correlations are low, the total model error can increase.

So there are obviously trade-offs between reducing the number of benchmark variables and getting results for more areas, and increasing the number of benchmark variables and getting more accurate and more generalisable weights. This trade-off has to be decided by the user.

10.1.3 The Representativeness of the Survey Data

For static spatial microsimulation models to work, the survey data being used must be similar to the areas being benchmarked to; otherwise, the reweighting procedure will not be able to find a population from the survey that represents the benchmark tables.

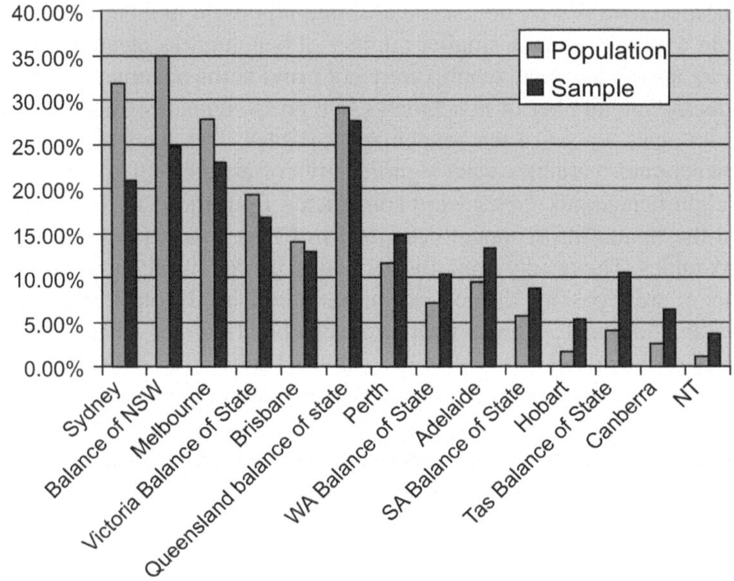

Fig. 10.1 Population proportions and samples, 2002–2003 and 2003–2004 surveys of income and housing (Source: 2002/03 and 2003/04 survey of income and housing (ABS 2004, 2005))

For some authors, this has meant that for the reweighting, they only use survey data from the broad region being estimated. This is the technique used by Anderson (2007), which gives a weight of 0 to anyone not in the area, so they are discounted from the reweighting. It is described as 'not filling Sheffield with Londoners' (Anderson 2007, p. 15) and is called by Williamson in Chap. 3 stratified household selection. The problem with this method is that it excludes what may be good records that are not in the area. Ballas extended this by deriving a geographical weighting technique (Ballas et al. 2005) which increased the chances of local households being used, but still allowed out-of-area households. This technique used a geographical constraint table as a benchmark, along with the other benchmark tables.

Other spatial microsimulation authors also permit households from any area to be included as long as they match the constraining criteria (Procter et al. 2008; Tanton et al. 2011), although this choice is all down to the initial data preparation and it is possible to limit the models to only area-specific residents if required.

Given that the Australian Bureau of Statistics also uses a reweighting method to reweight their survey data, it is interesting to look at the distribution of sample records across Australian state and territories. This will give some idea of where people come from who are then being used to derive state-level estimates from Australian survey data.

Figure 10.1 shows, for the Australian Survey of Income and Housing in 2003–2004, what proportion of people in the survey come from each area in Australia. The areas are capital cities in each state and balance of state in most states. The ACT does not

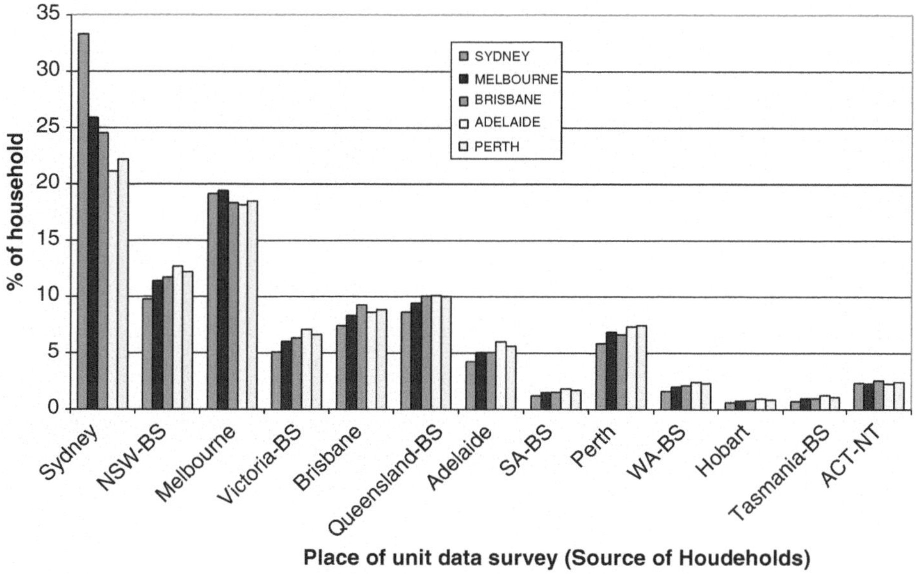

Fig. 10.2 Source of households to populate SLAs in SpatialMSM/08c in five capital cities (Source: SpatialMSM/08c applied to 2002/03 and 2003/04 survey of income and housing (ABS 2004, 2005))

have a balance of state category, and the Northern Territory does not have a capital city category.

In Fig. 10.1, the population bar is the proportion of the Australian population in that area (33 % for Sydney), and the sample bar is what proportion of the ABS sample came from that area (22 % for Sydney). It can be seen that the major capital cities in Australia (Sydney, Melbourne and Brisbane) are under-represented in the ABS samples, and the smaller states and capital cities in Australia (WA, SA, Tasmania and the ACT) are over-represented. This oversampling in smaller states is intentional on the part of the ABS (ABS 2002).

An obvious question to then ask is how a reweighting algorithm uses households from different states to represent SLAs in each state. Figure 10.2 shows the proportion of households used to populate SLAs for capital cities and non-capital cities (Balance of State, shown as BS in Fig. 10.2) in Australia using the SpatialMSM spatial microsimulation model described in Chap. 6. While other microsimulation models would give different results, the results from this model are illustrative.

So the first bar in Fig. 10.2 shows that 33 % of Sydney households, 10 % of NSW non-Sydney households, 19 % of Melbourne households and so on (all adding to 100 %) were used to populate Sydney SLAs.

It can be seen from Fig. 10.2 that Sydney households are used to estimate the SLAs of the majority of other capital cities (e.g. 25 % of the households in the Melbourne model were Sydney-based households in the survey dataset), whereas the Hobart households are not used much at all. The reason for this is that there are so few of them in the sample – so while Fig. 10.1 shows that Hobart households are

Table 10.1 Effect of using households from each capital city to estimate areas in the capital city using spatial microsimulation

Source of data for estimation with SPATIALMSM/08c (11BM)	Number of SLAs which passed the accuracy criteria	Number of SLAs which failed the accuracy criteria	SEI
Sydney for Sydney	63	1	0.9676
Australia for Sydney	63	1	0.9618
Melbourne for Melbourne	78	1	0.9263
Australia for Melbourne	79	0	0.9511
Brisbane for Brisbane	214	1	0.9263
Australia for Brisbane	212	3	0.9224
Adelaide for Adelaide	55	0	0.9735
Australia for Adelaide	55	0	0.9534
Perth for Perth	35	2	0.8478
Australia for Perth	35	2	0.7856

Source: SpatialMSM/08c applied to 2002/03 and 2003/04 survey of income and housing (ABS 2004, 2005)

over-represented in the sample, there are still only 731 of them in the two samples used for the SpatialMSM model, compared to 2,862 Sydney households.

The next step in this process was to test whether better results for small areas in each Australian capital city were obtained from the spatial microsimulation model if it was run with households from that capital city only. To do this, a number of subsets of the original sample were selected with all households in each capital city in each Australian state, and the spatial microsimulation modelling for small areas in that city was done using this subset of households only. The number of small areas which passed the accuracy criteria and the accuracy of the overall estimates were then calculated. The results are shown in Table 10.1.

It can be seen that there was not much difference between the two approaches in terms of the number of small areas that passed the accuracy criteria; however in terms of the SEI, using Perth households to estimate small areas in Perth gave much better results than using all households to estimate small areas in Perth. For all other capital cities, using the households in that capital city gave slightly better results, except for Melbourne where they were worse.

Further work using the SpatialMSM model has also shown that small areas in Tasmania are estimated more accurately using households from Tasmania only, and we suspect there may be a number of reasons for this (Vidyattama and Tanton 2011). One may be that there is a more homogenous population in these areas; so where the population is more homogenous, better estimates are obtained using households from that area. If the population in an area is heterogeneous, then using a wider variety of households to estimate the area should provide better estimates. Another explanation may be that the survey being reweighted contains a much higher number of people from New South Wales (1,958 households from NSW compared to 670 households from Sydney), which will affect the estimation process for Tasmania as people from Sydney are very different to people from Tasmania (higher incomes, different age structure, etc.). In terms of proportions, Tasmania has a much higher representation in the sample (670 out of 202,400 households is 0.33 % compared to 1,958 out of 2,651,700 households or 0.07 % in NSW – see ABS 2009). What is important for the SpatialMSM model is the number of people being used to fill an area, and this means NSW has a much higher number.

So overall using households from the area being estimated will give slightly better results with little difference in the number of SLAs which failed the accuracy criteria. The fact that the accuracy has increased more in the two most unpopulated capital cities (Adelaide and Perth) shows that the Australian sample for these two cities may give a slightly misleading result since the sample is dominated by households from the larger capital cities (see Fig. 10.1).

10.2 Conclusions

This chapter has looked at a number of issues when conducting static spatial microsimulation modelling. These issues have included data limitations, working out the optimum number of benchmarks, and identifying how representative the survey data are.

What we find is that the main limitation in the data is getting two comparable datasets, a survey and census dataset. There may need to be some adjustments to get matching data definitions and categories.

In terms of the number of benchmarks, increasing the number of benchmarks makes the final weights more generalisable but increases the number of non-converging areas. On the other hand, having too few benchmarks means the results are not accurate, so it is important to validate the final estimates to ensure enough benchmarks have been used.

In terms of using records from a sample that are not from the area being estimated, we find using within-area records may give slightly better results for some areas, but not for all areas. So while there may be some slight advantage in using a same-city sample for the smaller capital cities, generally using all records in the sample gave better results. The advice would be that when deriving estimates for broad areas that were not sampled well in the original survey, use records from that broad area; but if it is a broad area that was sampled well, then using all observations will give a better result.

References

ABS. (2002). *Information paper: Labour force sample design* (Cat # 6269.0). Canberra: ABS.

ABS. (2004). *Household income and income distribution 2002–03* (Cat # 6523.0). Canberra: ABS.

ABS. (2005). *Household income and income distribution 2003–04* (Cat # 6523.0). Canberra: ABS.

ABS. (2009). *Household income and income distribution 2007–08* (Cat # 6523.0). Canberra: ABS.

Anderson, B. (2007). Creating small-area income estimates: Spatial microsimulation modelling. *Communities*. http://www.communities.gov.uk/documents/communities/pdf/325286.pdf. Accessed Aug 2011.

Ballas, D., Clarke, G., Dorling, D., Eyre, H., Thomas, B., & Rossiter, D. (2005). SimBritain: A spatial microsimulation approach to population dynamics. *Population, Space and Place, 11*(1), 13–34. http://dx.doi.org/10.1002/psp.351

Birkin, M., & Clarke, M. (1988). SYNTHESIS – A synthetic spatial information system for urban and regional analysis: Methods and examples. *Environment and Planning A, 20*(12), 1645–1671.

Birkin, M., & Clarke, M. (1989). The generation of individual and household incomes at the small area level using synthesis. *Regional Studies: The Journal of the Regional Studies Association, 23*(6), 535–548.

Procter, K., Clarke, G., Ransley, J., & Cade, J. (2008). Micro-level analysis of childhood obesity, diet, physical activity, residential socio-economic and social capital variables: Where are the obesogenic environments in Leeds? *Area, 40*(3), 323–340.

Tanton, R., & Vidyattama, Y. (2010). Pushing it to the edge: Extending generalised regression as a spatial microsimulation method. *International Journal of Microsimulation, 3*(2), 23–33.

Tanton, R., Vidyattama, Y., Nepal, B., & McNamara, J. (2011). Small area estimation using a reweighting algorithm. *Journal of the Royal Statistics Society Series A*. doi:10.1111/j.1467-985X.2011.00690.x

Vidyattama, Y., & Tanton, R. (2011). Estimating the spatial distribution and characteristics of households in Tasmania. Final Report, Tasmanian Department of Premier & Cabinet: Hobart.

Part III
Dynamic Spatial Microsimulation Models

Chapter 11
Moses: A Dynamic Spatial Microsimulation Model for Demographic Planning

Belinda Wu and Mark Birkin

11.1 Introduction

Due to its vital role in human society, the study of population has always been at the centre of public policy and planning. People's movements, interactions and behaviours will inevitably have an important impact on the society and environment that they are living in. At the same time, changes in these factors will also lead to an evolution of the population itself over time. As advances in technologies and new tools often bring new visions, computer-based models have now been extensively used in modelling complex social systems. This is not only because they can provide valuable groundwork when it is too expensive or impossible for practical reasons to experiment in reality, but also new research methods enabled by the capabilities of modern computers can radically transform human ability to reason systematically about complex social systems. This has become increasingly important as our world today confronts rapid and potentially profound transitions driven by social, economic, environmental and technological changes.

To facilitate strategic decision making and to plan developments for our future, it is vital to study and understand changes in our population. Traditionally, macro-simulation has been used to model populations. However, macroscopic simulations have limits in representing small-scale (or microscopic) occurrences, discontinuity and heterogeneity within a system. This limits their effectiveness in studies where individual characteristics are important. On the other hand, the microscopic approach can deal with the rich details of individuals. Microsimulation provides insight into the behaviour of a system under a range of conditions. With the capability to provide valuable information over a wide range of individual inputs, microsimulation models

B. Wu (✉) • M. Birkin
School of Geography, University of Leeds, Leeds, UK
e-mail: Belinda_Wu25@hotmail.com; geombw@leeds.ac.uk

R. Tanton and K.L. Edwards (eds.), *Spatial Microsimulation: A Reference Guide for Users*, Understanding Population Trends and Processes 6, DOI 10.1007/978-94-007-4623-7_11,

(MSMs) have become not only an excellent discovery tool but also an indispensable assistance to strategic decision making (Orcutt 1957; van Imhoff and Post 1998; Harding 2007).

Dynamic MSMs are not only able to update the characteristics of the micro-units caused by the stimulation of endogenous factors but can also project them over time to include demographic processes and social economic transitions, such as ageing, mortality, fertility or social and geographical mobility (O'Donoghue 2001). Therefore, they provide a better longitudinal representation of the studied population than static MSMs that traditionally use an "arithmetical calculator" approach (Harding 2007). At the same time, human activities have a strong spatial dimension. A spatial MSM takes the spatial dimension into consideration. Accordingly, the population can be simulated within a local context and can also pick up on regional characteristics that may be the outcome of multiple interwoven factors that cannot otherwise be easily modelled all together.

This chapter introduces Moses, a dynamic spatial MSM that dynamically simulates the UK population through discrete demographic processes at a fine spatial scale to capture the local characteristics for a duration of 30 years from 2001 to 2031. In the following sections, we will describe the modelling approach, the components of the model, followed by a discussion and analysis of results.

11.2 Modelling Approach

Moses is an individual-based model that simulates the UK population through discrete demographic processes at a fine spatial scale for 30 years from 2001 to 2031. The modelling method is grounded in a dynamic spatial MSM, where each individual is described with their particular attributes and behaviour within a local context at each simulation step. In this section, we will discuss the Moses modelling approach in detail.

11.2.1 Dynamic Microsimulation Model

Brown and Harding (2002) define the term of social modelling as "the representation of social phenomena and/or the simulation of social processes" and describe microsimulation modelling as "a pre-eminent type of social model". Modern social science studies now often require detailed information about the interactions between the policy and the socio-economic behaviours of individuals, and MSMs model such interactions through the simulation of distinctive behaviours and characteristics at the level of individual decision-making units. Advances in computing and analytical techniques now allow MSMs to portray with great sophistication answers to a range of questions that researchers may ask when modelling a large, complex social system at the level of individuals.

Static and dynamic approaches give rise to two rather different types of MSM. In what follows, some of the main differences are identified whilst recognising that the terms "static" and "dynamic" are sometimes used in different ways by various authors. Typically, a static MSM takes a large representative sample with detailed information and uses synthetic reweighting techniques (although other techniques are also used: see Part II of this book) to generate the demographic and economic characteristics expected at some future time point. Normally, static MSMs simulate only the immediate or "morning-after" policy impact upon individual decision units. If changes in demographic structures over time are required, with a static MSM, this would be performed by using static ageing techniques and by re-running the cross-sectional simulation using future population estimates at specific time points. Such static MSMs "have usually been arithmetical calculators" (Harding 2007) and normally simulate the change "under the assumption that individual behaviour is unchanged" (Bourguignon and Spadaro 2006). Because the change in the demographic structure of the modelled population is performed by reweighting using some external information, it focuses on what the external information brings to the population, and therefore, it does not model the changes in population itself. A typical "what-if" static MSM scenario would be the following: if there had been no poll tax in 1991, which communities would have benefited most and which would have paid more tax in other forms (Ballas et al. 2005; Harding 2007; Gilbert and Troitzsch 2005; Vidyattama et al. 2011; Chin et al. 2005)

In view of these limitations of static MSMs, dynamic MSMs have become increasingly popular in recent years. Dynamic MSMs use a technique where entities change their characteristics as a result of endogenous factors within the model. A certain degree of interaction between micro population units can be found in dynamic MSMs. Such interaction typically includes processes such as birth and marriage (O'Donoghue 2001). Dynamic MSMs rely on an accurate knowledge of the individuals and the dynamics of such interactions. Dynamic MSMs "try to move individuals forward through time, by updating each attribute for each micro-unit for each time interval" (Harding 2007). In a dynamic MSM, the typical updating of the demographic structure is performed by ageing the modelled population individually (by asking "yes or no" questions on important transitions such as birth, death, marriage, etc.) with transition probabilities according to life tables and/or exogenous time series. The changes in the population itself are modelled, and the simulation in one year may affect an individual unit's characteristics in the subsequent year. Thus, dynamic MSMs are particularly useful for longer-term "what-if" scenario explorations and projection purposes. A typical future-oriented "what-if" dynamic MSM scenario would be the following: if the current government had raised income taxes in 1997, what would the redistributive effects have been between different socio-economic groups and between central cities and their suburbs by 2011 (O'Donoghue 2001; Ballas et al. 2005)?

In Table 11.1, some of the most important differences between the static and dynamic MSMs are summarised. Although static models may be more effective at times for specific short-run projections and often demand less in computing resource and skills, dynamic MSMs feature more detailed and realistic population ageing

Table 11.1 Important differences between static and dynamic microsimulation models

Model feature	Type of MSM	Evaluation	Comment/discussion
Simulation procedure	Static	Deterministic/stochastic	Static approaches are largely deterministic as the population structure is fed into the model Individual population members are then simulated, albeit through a stochastic mechanism, in accordance with this structure
	Dynamic	Stochastic	Dynamic models are essentially stochastic, as the members of the population evolve over time through randomly generated transition processes
Ageing technique	Static	Static ageing	An artificial process: for example, if the population is getting older, this can be represented by an increasing preponderance of elderly individuals in the database, but there is no change in those individuals
	Dynamic	Dynamic ageing	Individuals are aged through model time so that someone aged 25 today becomes 26 after the first year of a dynamic simulation, and so on. Other characteristics can be adjusted dynamically in a similar way so that individuals could gain additional qualifications or accumulate equity in the housing market over a period of time
Entity interactions	Static	Not possible	Individual population members are considered as discrete and independent
	Dynamic	Possible (although not mandatory)	Interactions are allowed: for example, family relationships can be preserved through the simulation, or illnesses could be transmitted through personal contact
Time	Static	No time element	A cross-sectional view of population change is presented
		Change processes and events are built-in	Evolutionary processes – such as births, deaths, marriage and household change, as well as ageing – are all represented as state transitions within the dynamic model
Population change	Static	Homogeneous	There are no mechanisms to represent increasing heterogeneity within the population
	Dynamic	Diverse	The "emergence" of entirely new demographic groups could arise quite naturally. For example, ethnic minority populations in the UK are traditionally young and family oriented, but over time new structures such as single elderly or codependent adults could evolve dynamically within the model
Impact of previous step on next step	Static	No impact	Lack of interaction and temporal inflexibility implies that any time steps are mutually independent
	Dynamic	Significant impact	Impacts in one period are carried into the future. Thus, for example, if we are trying to simulate the effects of changing university fees on participation in higher education and its long-term influence on labour markets and economic competitiveness, then a dynamic MSM would likely be much more effective than a static one

and are also viewed as better at producing realistic long-term estimates, which account for interim changes in economic and demographic trends (O'Donoghue 2001). In view of its many advantages, dynamic microsimulation modelling techniques are used by Moses to include important demographic processes in the dynamic changes of the modelled population. Compared to a static MSM, the dynamic MSM of Moses:

- Builds in change over time
- Has processes for introducing population units into the system of interest and taking them out
- Has the capacity of reproduction (through birth) and elimination (through death) of individual population members

11.3 Spatial MSMs

Hägerstrand's space–time geography revolutionised the study of society. He pointed out that the spatial and temporal dimensions must be included into social studies, as "One cannot be at two places at the same time" (Hägerstrand 1985). His research has explored the conceptual basis for developments in spatial MSMs, which are distinguished from other types of MSMs with the capability to simulate virtual populations in given geographical areas (Hägerstrand 1985; Ballas et al. 2005). In a spatial MSM, local contexts can be taken into account when studying the characteristics of these populations. We often find certain demographic characteristics persist in some areas, but it is difficult to determine the exact cause and model the process. Often it is the outcome of multiple elements interacting with each other at the same time. Taking into consideration its spatial dimension, the population can be simulated within a local context and picks up on regional characteristics that may be the outcome of multiple interwoven factors that cannot be otherwise easily modelled together. Location provides a useful proxy variable for the simultaneous operation of many variables, such as socio-economic, ethnic, lifestyle and environmental variables, without introducing too much theoretical and practical difficulty. As long as we know the profiles of the different locations, we can capture some of the effects of demographic change.

On the other hand, geography has a vital role in affecting social progress and welfare. Given the nature of social systems, it would not be complete without considering the spatial impact in a policy MSM. When assessing the impact of the policy changes on individuals, many studies have identified that the outcomes do vary spatially (Birkin et al. 1996; Ballas and Clarke 2001; Wu and Hine 2003; Tanton et al. 2009). Indeed, there is a need to estimate the geographical impacts as well as the socio-economic impacts of policies. From a planning/policy point of view, "Means are to be employed somewhere" (De Man 1988). Essentially, people have to live in a local area, and they are affected by what goes on around them. Some studies also believe that social policies can be seen as alternatives to area-based policies, and in some instances, spatial impacts of social policies can even be validated through the respective impacts of area-based policy studies.

Although area-based policies have a geographical impact by definition, there has been very limited analysis of the spatial impacts of policies that were not designed to have a geographical impact. They suggest that spatial MSMs can also be used for the design of proactive geographically oriented social policies (Ballas et al. 2005). Spatial MSMs are concerned with the creation of large-scale datasets estimating the attributes of individuals within the study area and are used to analyse policy impacts on these micro-units (Birkin et al. 1996). Spatial MSMs therefore have advantages over other MSMs in exploration of spatial relationships and analysis of the spatial implications of policy scenarios. Another feature of spatial MSMs is that they allow data from various sources to be linked and patterns to be explored at different spatial scales with re-aggregation or disaggregation of the data. They also allow updating and projecting, which is of particular importance in forecasting future patterns (Ballas and Clarke 2001; Wu et al. 2008).

A spatial MSM can be either static or dynamic, but within a dynamic spatial MSM, both the characteristics of the individual and the context can change. Moses uses a dynamic spatial MSM that simulates the change of populations over time in small geographical areas at the intra-urban scale.

11.4 Description of Model

In this section, we will describe how the dynamic spatial MSM called Moses is constructed, including the details about the demographic processes, representation of the population within the system and the modelling method. The urban area of Leeds, a city with a population of approximately 730,000 (in 2001) in the north of England, is used for illustrative purposes throughout this chapter. However, this model has been generalised for the whole of the UK.

11.4.1 Data

The main datasets used in the probability calculation are:

- Census data: Individual Samples of Anonymised Records (ISAR) and Household Samples of Anonymised Records (HSAR) (national level)
- British Household Panel Survey (BHPS) data (national level)
- Vital statistics (ward level)
- Special Migration Statistics (ward level)
- ONS mid-year estimation (2001–2006) and sub-national projections (2007–2031) (sub-national level)
- Various tables in population trend (various levels)

- Various other census tables and life tables (various levels)
- Commissioned tables and bespoke data (various levels)

The baseline population is produced by a model that synthesises individual characteristics on a progressive basis using compound probabilities. The probabilities are calculated using the technique of iterative proportional fitting (IPF), as used in many other MSMs (Birkin et al. 2006). We already have a distribution of individuals (within the ISAR) that are not representative of each small area due to the sample size. So we need to select from the ISAR in order to define a subset which is more representative of the area. This iterative process can be thought as assigning weights of one or zero to each ISAR record until the appropriate record with suitable characteristics/attributes is selected (Williamson et al. 1998). ISAR provides over 100 individual attributes.

HSAR and BHPS data are then used to introduce further variables, such as the household and lifestyle attributes (Wu et al. 2008), using the same method. Vital statistics is used for ward level birth and death information. Various life tables have also been used in the mortality probability calculation. Special Migration Statistics have been used for the migration probability calculation. ONS mid-year estimation and sub-national projections and various tables in population trend have been used to update the population trends. There are also various other census tables, commissioned tables and bespoke data that have been used in various processes, for example, age at marriage.

11.4.2 Demographic Processes

Six important demographic processes have been modelled in Moses: ageing, mortality, fertility, household formation, health change and migration. Each component of change is simulated in separate modules, but individual components can also affect each other during the simulation. For instance, household formation will lead to migration in many cases. Mortality, fertility and migration are modelled simply because they are the three fundamental components of any population change. In this study, we also consider that knowledge about ageing, household formation and health is useful for public policymaking and demographic planning due to their impact on housing, transport, health and other public service provision. Furthermore, these processes can lead to changes in other demographic processes. For example, household formation has a key influence on fertility, especially when associated with a transition to marriage.

The demographic processes which are incorporated within Moses are illustrated in Fig. 11.1. Transition probabilities for each of these events are applied at discrete 1-year intervals. The rate of change for each of the components depends on both observed historical trends in the area and on forecast national trends. For more information about how these transition probabilities are calculated for each of the six demographic processes, see Sect. 11.4.4 below.

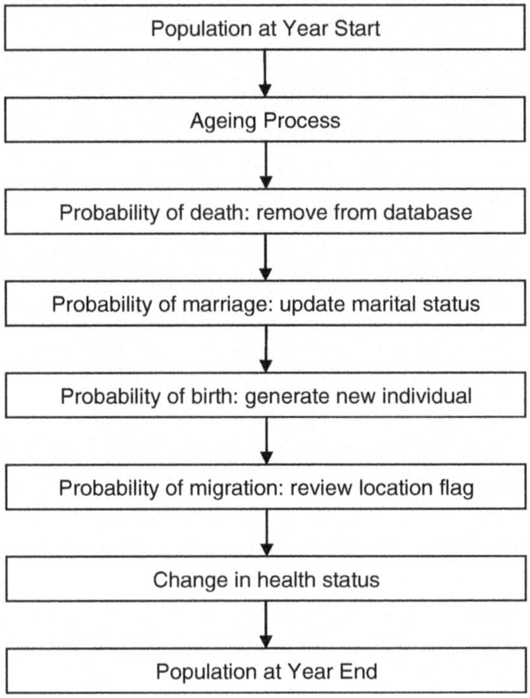

Fig. 11.1 Processes included in the population simulation

This sequence of events could be placed in many different orders. That said, it is considered more logical to evaluate fertility following the formation of marriages and partnerships. Similarly, mortality is considered early in the process for practical reasons since if an individual life course is terminated, subsequent processes can then be ignored.

Due to the nature of the demographic events, some processes are easier to model than others. Processes such as mortality and ageing are straightforward, whilst some processes are more complicated and may need to be modelled in multiple stages. For example, a migration process will require three stages of modelling to find out the answers for:

- Who to move (based on a probability of moving or staying)
- How to move (based on a probability of household or individual move)
- Where to move (based on a probability of moving to a certain area)

Different processes can also be interdependent, for example, marriage and migration processes are often connected, as a change in marital status will frequently occur alongside the move to a new home. This modularised design provides great flexibility in both model development and maintenance, as well as allowing the possibility of running the model with different combinations of demographic processes. For instance, if only the natural change of the population is of interest, the simulation of other processes can be "switched off", and only the processes of ageing, mortality and fertility can be used.

11.4.3 Representation of the Population in the System

The studied population is modelled as individuals in households (including single household) or communal establishments (formal care facility, prisons, army, etc.). The communal establishment populations are identified specially, and transitions for these people are not modelled. However, the total number is included in the baseline population in the initiation of the simulation. Each individual is also allocated to a specific small area within the study area.

Another feature of Moses is that it includes not only individuals but also the households through which they interact with the rest of the world, through interactions with other people and the environment that they live in. Thus, although the studied population is modelled as individuals, there is an interdependency between the household, individual and environment.

For instance, during the process of marriage, the formation of a new household between two individual households will mean changes in at least one individual's location. If it involves two households, this will result in changes in both households if one is going to join another in an existing household or even create a third household, and changes in both old households if both move out of existing households. The areas that they were/are going to live in will experience both local housing changes and local population changes. Similar changes will be experienced from any migration process. Due to this interdependency, the operation of these demographic processes of individuals also leads to the formation and dissolution of households during the simulation process.

11.4.4 Modelling Method

Moses dynamically simulates individual changes through the application of transition probabilities for each of the six discrete events at 1-year intervals. For instance, assume we start with a population of entities, set P, made up of individuals [P^1, P^2, … P^n] where n is the number of individuals in the population sample. Each individual has a set of attributes, [a_1^t, a_2^t, … a_m^t], which describe the individual at the time t. We therefore have an $n \times m$ array of person attributes. This array is populated with a synthetic population recreated from the census samples (for more details of the population recreation, see Birkin et al. 2006).

Then we update the population by applying transition probabilities to individual attributes for each simulation step so that the baseline population [$P^1_{a_1^t a_2^t \ldots a_m^t}$, $P^2_{a_1^t a_2^t \ldots a_m^t}$, … $P^n_{a_1^t a_2^t \ldots a_m^t}$] changes to new sets with attributes/states at a point in time $t+1$, $t+2$, … and so on. Each end population for a given year then becomes the start population for the next simulation year. Therefore, the impacts of previous changes have been taken into account for the next year's simulation.

Transition probabilities for each of the six demographic processes are strongly influenced by both age and sex (Rowland 2003). They also vary by geographical area as a result of the differences in social, economic and environmental profile of area populations. Therefore, the location of individuals can provide the environmental context as well as provide some impact from other relevant factors that are not modelled specifically in this model. Accordingly, the Moses model assumes that all the demographic processes modelled are the functions of age, gender (except the fertility process where only women are concerned) and location. Therefore, a change/an event occurs to an individual in a demographic process if

$$\text{Ran}(0,1) \le P(k_2 | a, s, l, k_1), \tag{11.1}$$

where Ran is a random number between 0 and 1, a is the age, s is the sex and l is the location of the individual and $P(k_2|a,s,l,k_1)$ is the probability of the event occurrence for an individual of age a, sex s at location l with current characteristic of k_1 to change to a characteristic k_2. For instance, in mortality, k_1 might be "alive", and $k_2$will be "dead"; in fertility, k_1 might be "no child born alive during the year", and k_2 will be "child(ren) born alive during the year".

Transition probabilities for each of the events within the six demographic processes are applied at discrete 1-year intervals. According to the nature of specific demographic process, a slight variance has to be introduced in calculating the probabilities. For instance, due to the nature of the fertility process, it only considers women at risk by age, marital status and locations.

In the attempt to capture the local context, we tried to calculate localised transition probabilities for two of the six demographic changes, whilst other processes used more aggregated probabilities for the whole of England. A method has been developed to localise the probabilities for mortality and fertility processes to the ward level using selected national, sub-national and ward-based datasets on the basis of the method explained by Rees et al. (2004). In the mortality process, transitional probabilities for individuals in each of 33 Leeds wards have been calculated using all three levels of data, considering both sexes and 101 single-year age bands (from 0 to 100+ years). For this single urban area, there are 6,666 mortality probabilities in total (this is calculated as 101 age bands; for both sexes, $2 \times 101 = 202$; for 33 wards in Leeds, $33 \times 202 = 6{,}666$).

For example, the survival probability for the Leeds population in 2001 takes the form shown in Table 11.2, which summarises the 6,666 different transitional probabilities for the mortality process (the survival probabilities, which are the opposite of mortality probabilities in the mortality process, are used in our model as only the survivors will remain in the rest of the simulation). Similarly, the fertility probabilities are based on single year of age and the marital status of the mothers. This produces 2,112 localised probabilities (age under 15–45+, 32 age bands; for marital status of single or married, $2 \times 32 = 64$; 33 wards in Leeds, $64 \times 33 = 2{,}112$).

Monte Carlo simulation is used as the main simulation method. The reason it was used was because it is useful when it is infeasible or impossible to compute an exact result with a deterministic algorithm. The Monte Carlo method converts uncertainties about the relationship between input variables and output variables of a model into conditional probabilities. By combining the distributions and randomly selecting values from them, it recalculates the simulated model repeatedly and brings out the

Table 11.2 Survival probability example: Leeds in 2001

Male							Female						
Area													
Age	Ward 1	Ward 2	Ward 3	…	Ward 32	Ward 33	Age	Ward 1	Ward 2	Ward 3	…	Ward 32	Ward 33
0	1	0.99018	0.992677	…	0.996948	1	0	1	0.99766	0.996945	…	1	0.994098
1	0.999547	0.997993	0.99918	…	0.998998	1	1	1	0.999145	0.999659	…	1	0.998759
2	0.999263	0.998526	1	…	0.998923	1	2	1	0.999032	1	…	1	0.999053
3	0.99951	0.999021	1	…	0.999284	1	3	1	0.999357	1	…	1	0.999371
4	0.999631	0.999262	1	…	0.99946	1	4	1	0.999515	1	…	1	0.999526
5	0.999729	0.99945	0.999765	…	0.999762	1	5	1	0.999786	1	…	1	0.99979
6	0.99981	0.999606	0.999585	…	1	1	6	1	1	1	…	1	1
7	0.99982	0.999628	0.999608	…	1	1	7	1	1	1	…	1	1
8	0.999821	0.999629	0.99961	…	1	1	8	1	1	1	…	1	1
…	…	…	…	…	…	…	…	…	…	…	…	…	…
92	0.821179	0.900588	0.878387	…	0.804951	0.752563	92	0.806547	0.834905	0.805754	…	0.819787	0.852985
93	0.807478	0.892666	0.8683	…	0.790129	0.734257	93	0.791834	0.822168	0.790987	…	0.805989	0.84154
94	0.793715	0.884524	0.859009	…	0.775312	0.716253	94	0.77712	0.809323	0.776222	…	0.792135	0.829937
95	0.77785	0.875175	0.847755	…	0.758215	0.695403	95	0.760142	0.794525	0.759185	…	0.776162	0.816582
96	0.766467	0.868564	0.839766	…	0.74591	0.680239	96	0.747927	0.783936	0.746925	…	0.7647	0.807059
97	0.755884	0.862066	0.832031	…	0.734605	0.666853	97	0.736692	0.773993	0.735655	…	0.754053	0.798
98	0.737503	0.850864	0.818672	…	0.714934	0.643404	98	0.717145	0.756748	0.716047	…	0.735559	0.782318
99	0.729555	0.846771	0.81355	…	0.706142	0.631764	99	0.708436	0.7495	0.707296	…	0.72754	0.775971
100	0.736705	0.752057	0.748081	…	0.73314	0.720414	100	0.733498	0.739597	0.73332	…	0.736405	0.743244

Source: Author's computation using ONS (2001)

Fig. 11.2 General simulation method

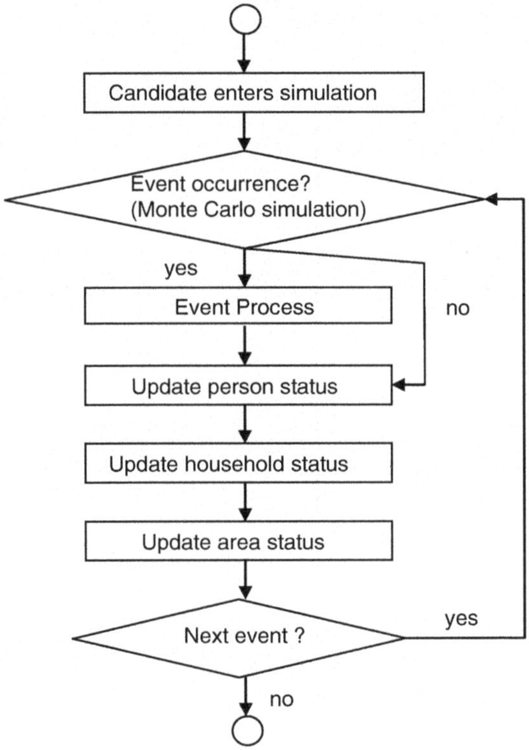

probability of the output. Therefore, each determination of whether a transition/event happens or not requires a new random number and a corresponding transitional probability for a person with certain attributes in that year. All probabilities are updated annually by applying weights based on the trend revealed in relevant ONS projections, and any change in any factor will result in the change of the probability and in turn the simulation process and result. The general microsimulation process used in every demographic transition in Moses uses the Monte Carlo method which is illustrated in Fig. 11.2.

11.4.5 Validation

Dynamic spatial MSM is hard to validate, due to the level of detail modelled in such MSMs. Often there are no appropriate microdata available that can be used to validate such details. However, the individual-based results can be easily re-aggregated to any spatial level to compare with other aggregated projections. In recent years, the ability to align the micro output to benchmark macro estimates has emerged as a crucial component of many MSMs, as alignment can help capture the macroscopic impact in microsimulation aggregate results and provide an indicator of the aggregate performance of the model (Rephann 2001). Typically this is achieved through adjusting the

Table 11.3 Result alignment analysis

Model	ONS			Moses		
Year	2002	2007	2031	2002	2007	2031
Total population	720,000	759,400	974,300	729,101	770,290	988,151
Natural change	1,000	3,000	5,500	403	3,929	1,506
Births	8,000	9,500	11,700	8,046	10,429	10,505
Deaths	7,000	6,400	6,200	7,643	6,500	8,999
Net migration	3,500	8,100	2,400	3,336	8,433	2,430
Internal in	28,000	30,500	30,500	28,354	30,939	30,932
Internal out	29,700	31,100	36,500	30,076	31,546	37,019
Immigration	9,800	13,500	16,500	9,925	13,696	16,733
Emigration	4,600	4,800	8,100	4,867	4,656	8,216

Note: For details of ONS data source, see Sect. 11.4.1

results from a model to reflect new individual outcomes, totals and flows (O'Donoghue 2001; Anderson 1997). Moses uses a naively disaggregation model to align to the UK official projections for Leeds, produced by the ONS model.

The aggregate probabilities used in the ONS model are naively distributed to the small areas and simulated under the same assumptions and trend information as revealed in ONS projections. The alignment results are then aggregated to the level of Leeds to find out if the dynamic spatial MSM can produce results that are consistent with the ONS projections. Although the influence of spatial variances is still visible in the aligned results, especially in fertility and mortality, those results indicate consistent patterns as the official projection in the total population, as well as in all components of change. Results of 3 years, 2002, 2007 and 2031, are presented in Table 11.3 to provide information on the alignment results at the start, middle and end of the simulation.

11.5 Model Result Analyses

The population are simulated through the six demographic processes from 2001 to 2031, and analyses have then been conducted on the simulated results. The main findings from the analyses are discussed in this section. As described before, the local authority (LA) area of Leeds is used for illustrative purposes. The results will be discussed at both the aggregate level of LA and the disaggregated level of wards. Results are also analysed by demographic changes to provide an insight of the components of changes.

11.5.1 Population Patterns at the Level of Local Authority Area

Using dynamic MSM, the Moses model can provide projections of an individual-based local population with dynamic changes that truly reflect the demographic changes in a way that they would happen in the real world. In Moses, the populations are also simulated within small areas of wards, and the spatial variances in these small areas can be captured. This feature enables us to study the population with a local context. Such simulated results with a local context can then be summed to higher levels of aggregation to facilitate various study requirements. On a more aggregated spatial scale, they are particularly useful in providing indications of various population trends to facilitate strategic decisions or to explore different scenarios. In the following section, we will provide examples from three different aspects of the population changes at the LA level of Leeds. We then discuss the potential demographic planning applications of Moses through different uses of the simulation results at the LA level.

11.5.1.1 Age–Sex Structure of the Population at the Level of Local Authority Area

Age and sex are the fundamental factors for population changes. Analysis of the age composition of populations is essential in demographic studies. The intensity of each of the demographic processes varies significantly by age but in different ways. At the aggregate level, demographers are interested in the age compositions of populations and the ages when people engage in certain behaviours. Due to biological factors, behaviour and well-being differences, males and females demonstrate considerable variance in their behaviours even when they are at the same age. In fact, the age–sex composition is so important to the nature and functioning of societies that all traditional population models are based on it (Rowland 2003). The population pyramid is a popular graphical device which can be used to illustrate populations' age–sex structures. In a population pyramid, the numbers/percentages of males and females in each age group are represented in the graph. As Moses outputs records of individuals with a rich set of attributes, population pyramids can be easily generated. Using the outputs each year, the changes of the age–sex structures in the study area can then be analysed over time.

At the local authority (LA) level of Leeds, population pyramids have been produced to illustrate the age–sex distribution of the Leeds population in 2001 and 2030 to demonstrate the population changes over time (Fig. 11.3). The darker shade represents the age–sex distribution in the year 2001, and the lighter shade represents the age–sex distribution in the year 2030. Here, we can clearly see how the Leeds population evolves over 30 years. The age–sex structure indicates a steady growth of the Leeds population. The largest growth has been seen in the age bands between ages 20 and 69, especially around the 40- and 50-year age groups. However, we also see a substantial increase in the ages 80–100+. The proportion of the population over 80 has almost quadrupled from 2001 to 2030. This indicates that ageing is an important trend in the Leeds population. The age–sex structure also indicates that there are more women in the age range 80–89, but the difference gradually reduces in the more elderly cohorts.

Using annual results from Moses, the trend of the population changes year by year can be monitored. Through observing the evolution in age–sex structures over

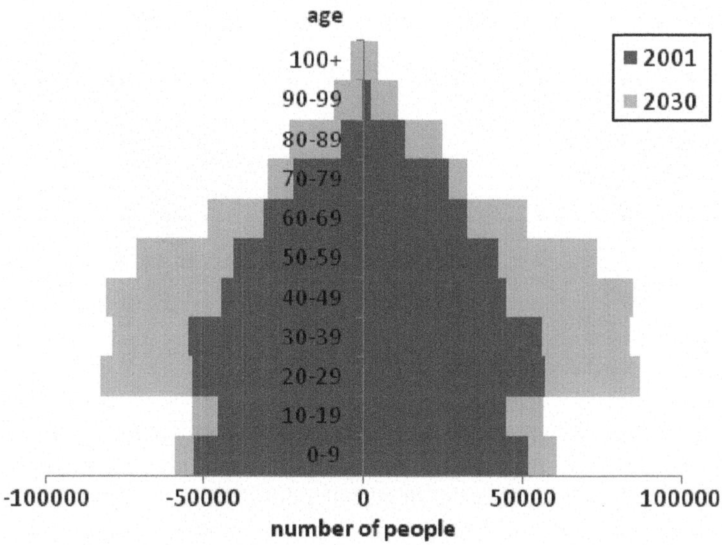

Fig. 11.3 The age–sex structure projections of Leeds population in 2001 and 2030

time, various demographic trends can be revealed. Such findings can then provide the groundwork for strategic demographic planning. For instance, the trends revealed from such analysis can be used to assist the development of early interventions. Different scenarios can be developed to explore the "what-if" situations. Such information can then be used to assess the impact of such demographic changes on various public plannings. For instance, it can be used to assess the requirement for public services provision, such as planning for public health services and transport services for an ageing society.

11.5.1.2 Sub-population Patterns at the Level of Local Authority Area

Moses can provide useful information on the Leeds population to facilitate strategic demographic planning in the area. However, sometimes we need to look further into patterns in different sub-populations to fully comprehend the changes in the whole population. For instance, the university student population has an important impact on the population structure of Leeds. Due to its central geographical location in the UK and its reputation for university education, Leeds has been attracting a large number of students to study in the local universities. However, students have a distinctive migration pattern. University students are highly mobile during their study. The first year students tend to stay in university accommodation. Later when they are more familiar with the area, they move out and find privately rented accommodation, often in the same areas that their fellow students would stay. They also move more frequently than other residents in the area, due to renting contracts and the annual summer break. However, the majority of them tend to stay in areas that are

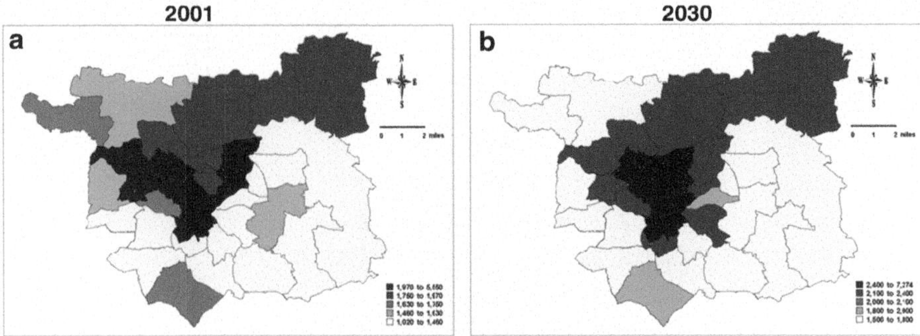

Fig. 11.4 Student population projections in 2001 and 2030

close to the universities where they study, which are usually in the centre of the city. Most of them will leave the area when they finish their study, for example, to take up employment elsewhere. Because of such features, student migration is an important component of Leeds migration, and we cannot understand migration in Leeds properly without looking into the details of this sub-population.

Moses allows us to focus on the student sub-population by using relevant characteristics (e.g. age and full-time student status) to identify students. Here, we present the projected university student population distribution in Leeds in years 2001 and 2030 in Fig. 11.4. The depth of the colour in the map indicates the density of the student population within the area (darker areas equate to more students). As we can clearly see, most of the university students in Leeds live in the city centre close to the universities. However, the two maps suggest that the students seem to move out of the areas in the northwest and south of the city and move into adjacent areas that are closer to the city. Thus, there is an indication in the map for 2030 that students will move closer to the city centre and that there will be a much more concentrated student population in the centre area.

Such findings are useful in understanding the impact of student migration on local migration. Similar analyses can be carried out on any sub-populations, and such studies help us understand the local population trends where such sub-populations have a significant impact. In turn, they provide a better groundwork for various demographic planning processes. The student migration patterns can be further explored in different scenarios to assess the impact on the Leeds population in "what-if" situations, for example, what changes will be brought to the population structure change in year 2030 if there is a dramatic increase in the number of student migrants into Leeds from 2010?

11.5.1.3 Demographic Changes at the Level of Local Authority Area

Changes in individual demographic processes also play an important role in understanding changes in the whole population. Moses can provide this information through the output of simulated population results by individual demographic

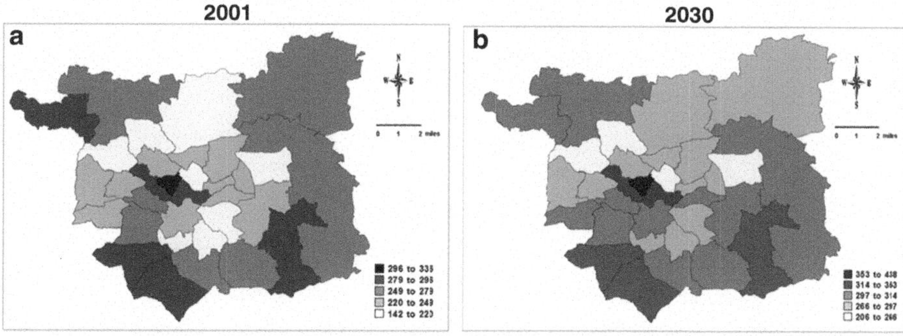

Fig. 11.5 Mortality projections of Leeds in 2001 and 2030 generated by Moses

processes. Such results are individual based and can be easily aggregated for various planning interests. As mortality is one of the most important components of change in a population, Fig. 11.5 provides the map of the mortality distribution of Leeds in years 2001 and 2030 to compare the changes in Leeds mortality over time.

In the maps, we can see that overall, the mortality analysis indicates that there may be some improvement over the years. A couple of suburban areas in the north and south seem to experience a particularly good reduction of deaths in the area. However, certain areas seem to experience a higher mortality than in 2001. This may indicate that there is an elderly or ageing population whose overall impact offsets mortality improvement over the years. Overall, the projection map indicates that the north of the city has seen more improvement in mortality than the south and the mortality remains high in city centre area. This may suggest the impact of migration, as the northern areas of the city are more established suburban areas which attract the healthiest or more affluent migrants, whereas migrants seeking work and from ethnic minority groups, perhaps with lower life expectancies, tend to move into the south and the centre areas of the city (Fig. 11.5). Such changes in mortality have an important impact on the population structures. Similar analyses can be conducted on any demographic process, and such analyses on the LA level can provide vital information for strategic planning of the city.

11.5.2 Results: Analysis at Small Area Level of Wards

The aggregated results of the microsimulation model are useful in facilitating strategic planning or policymaking. However, at an operational level, local context plays an indispensible role in area-based intervention measures. Therefore, this chapter also assesses the Moses results by small areas to explore the spatial differences. Cookridge and Headingley are the two wards that have been selected as examples to demonstrate the spatial variance in small areas of Leeds. They are selected because of their distinctive local characteristics. Cookridge is an established suburban area in

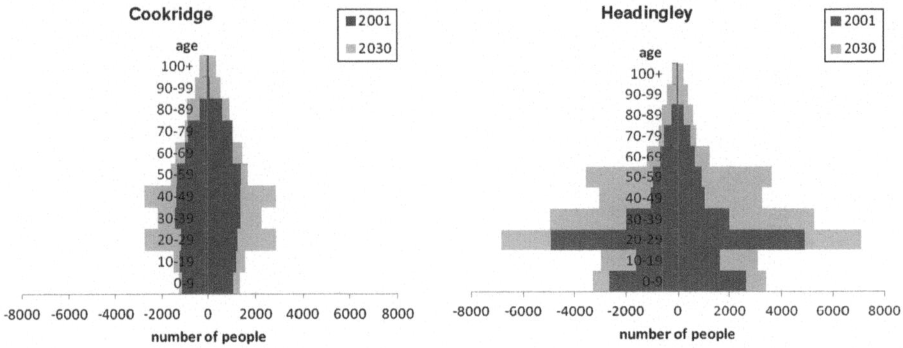

Fig. 11.6 The age–sex structure of ward population: Cookridge and Headingley in 2001 and 2030

the north of Leeds, whilst Headingley is an area close to the city centre, where many university students and young professionals live.

In the following section, we will assess the spatial variance in the local populations in the two wards through analyses of the age–sex structure, sub-population patterns and demographic changes.

11.5.3 Age–Sex Structure of the Populations at the Small Area Level

The simulation results over 30 years for all 33 wards of Leeds have been analysed, and a substantial difference is revealed between the small area projections of Cookridge and Headingley. In Fig. 11.6, the population changes over time in the two wards have been presented in population pyramids. As before, the darker shade represents results in year 2001, and the lighter shade represents results in year 2030.

From Fig. 11.6, we can clearly see the different population characteristics in small areas. Headingley is an area where many students live during their university studies. Students tend to leave the area upon the completion of their studies, and new students move into the area. Due to the replenishment of the student population, the population in Headingley stays younger than the rest of the Leeds population. There are more young people, especially aged 20–29, in Headingley than in any other small areas. In contrast, Cookridge is a more established suburban area where the local population ages more obviously than Headingley. There is a substantial increase in population aged 90+ in Cookridge. Headingley also sees changes in this age group in 2030, but these are much less obvious than Cookridge. Those aged 20–29 stay as the largest group of the local population, although there has been a substantial increase in the number aged 30–60 in 2030. The population in Headingley also grows much faster than that in Cookridge.

From the age pyramids for these small areas illustrated in Fig. 11.6, we can clearly see that characteristics of the local population evolve differently in small areas. The replenishment of the students and a younger composition of the population allow the Headingley population to stay younger and keep growing. From the projections in 2030, we can see a larger cohort of older people in the small area populations such as Cookridge than in areas such as Headingley. Such information is very important for both demographic and public planning. For instance, it will assist the planning of the health service provision in areas such as Cookridge which are experiencing an increase in the number of older people.

11.5.3.1 Sub-populations at the Small Area Level

As described in Sect. 11.5.1.2, analyses of sub-populations are useful for understanding the changes on the whole studied population and for providing the groundwork for strategic planning. The analysis of the migration patterns of university students in Leeds using the simulation results has demonstrated the usefulness of the analysis of simulation results of this sub-population.

Such findings can be useful to understand the impact of student migration on the local migration, not only in the city of Leeds as a whole but also in small areas. In fact, it may be the most important change in some small areas where student migration is high. Therefore, such analysis can provide important insights for understanding the local population trends. Such insights can then allow planners to develop location-based plans or policies in order to appropriately support different areas.

The patterns revealed in the maps in Fig. 11.4 can be used for other planning purposes as well. For instance, such information can be used to assist the planning of student housing and even policing (e.g. to pinpoint the hot spots for antisocial behaviour fuelled by alcohol consumption during the weekend and burglaries related to student accommodation during university holiday time). Moses can provide information on sub-populations in each small area to meet the need of such area-based investigations and planning.

11.5.3.2 Demographic Changes at the Small Area Level

Compared to other types of demographic models, Moses can also provide information about the spatial variances in different demographic processes. Headingley and Cookridge have again been used here for demonstration purpose. As previously described, the two local populations have considerable differences.

In this section, we will look into the differences in mortality between these small areas using the results from Moses. The simulation results from the mortality module are shown in Fig. 11.7. The initial results suggest that Headingley may have a lower mortality rate than Cookridge in 2001, except for the over 100 age group (this may be caused by the small number of people in the age band of 100+, e.g. when there is only one person aged 100+ in the area, his/her death will lead to a 100% mortality

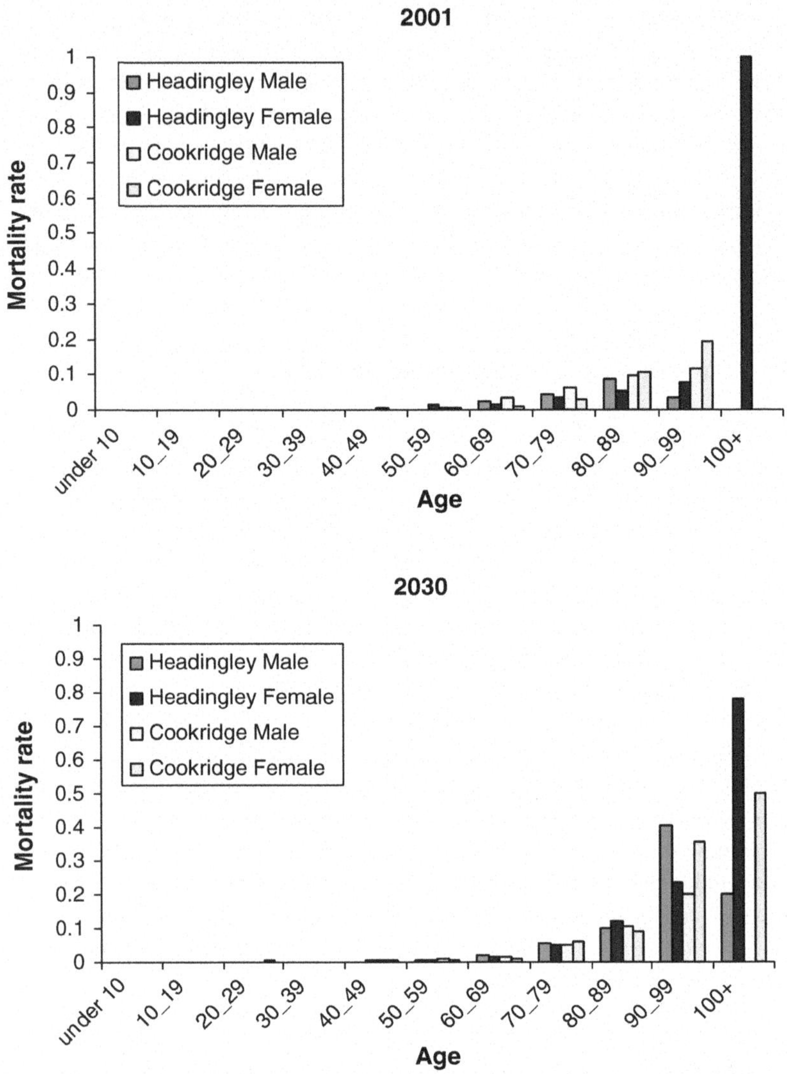

Fig. 11.7 Projections of mortality in small areas: 2001 and 2030

for this age band). However, the mortality rates then escalate in Headingley, and they are substantially higher than in Cookridge in 2030, especially for older ages (90+ and 100+). This may indicate that Cookridge experiences more improvement in mortality than Headingley. However, for the very old age groups, there are fewer older people aged 90+ in Headingley, and a single death can lead to substantially higher mortality rates for this age band (Fig. 11.7).

Such analysis reveals the change in the demographic processes in small areas which underlie the population structure changes. These patterns will in turn lead to changes at the more aggregate level. Moses enables us to understand the aggregate population changes from the underlying small area differences in each demographic process. Such understanding can play an important role in demographic planning or other relevant strategic decision making.

11.6 Conclusions

The Moses dynamic spatial MSM provides the characteristics of the studied population for individuals through a truly dynamic ageing process. A rich set of attributes can be updated through the evolution of demographic changes and interactions of the demographic processes. Individuals are modelled within small areas (wards) to reflect the local context. At each simulation step, attributes of each individual are updated through Monte Carlo simulations using transitional probabilities that are calculated using relevant demographic and spatial information in an attempt to capture their demographic characteristics and local area characteristics. Such changes are then built into the baseline population for the next simulation step. As all changes are dynamically simulated each year and driven by multi-criteria-based probabilities, including local area factors, the overall results present a much more robust representation of the studied population compared to static models or to aggregate models that overlook the spatial variance.

The dynamic spatial MSM features enable Moses to produce better projections of changes in the baseline population. Moses also allows exploration and analysis of various scenarios on the population by area, sub-population or demographic process. Using Leeds as an example, Moses has demonstrated its strength in providing a better representation of a studied population and provides an assessment of multiple scenarios for different planning applications or social futures. For example, in current work, the effects of an "epidemic" in obesity on health status and life expectancy are under exploration. Deteriorating personal health through obesity could easily mitigate or even reverse the improvements in life expectancy and demographic expansion, which have been illustrated elsewhere in this chapter. Equally, specific local policies, especially relating to transport and housing, will influence local patterns of growth and structural change in the population (see, e.g. Birkin et al. 2010).

At the aggregate level of Leeds, the results from Moses can demonstrate the trends in population year by year. For more details, we can even trace various characteristics of sub-populations or even individual demographic processes over a long period of time. This will not only help in understanding the underlying changes as a population and community evolves but also provides useful insights for strategic decision making. At the small area level of wards, it is found that characteristics of the local population changes differently. The small area analysis confirmed that population evolution does vary at a fine spatial scale. Moses allows us to study the spatial variance in age–sex structures and in sub-population characteristics, as well as in different demographic processes. Such disaggregated dynamic changes in the studied population provide useful insights into understanding population patterns at a more aggregated spatial level. At the same time, such disaggregated explorations provide valuable information for tactical decisions and location-based studies and policies.

As a demographic planning tool, Moses can monitor the evolution of population structures and various demographic changes at a fine geographic scale. This provides vital information for demographic planning/policymaking (especially location-based policies). Moses can also benefit other public policymaking or public service planning. For instance, the ageing trends in certain suburban areas may promote changes in health service and public transport service provision in order to enable easy access to such services for the old and frail in the area. The attributes captured in the system are also very useful for different policy analyses and research.

The Moses model has provided a framework to enable the effective modelling of heterogeneous decision-making units on a large scale. The model itself provides a useful tool in assisting decision making, exploring various "what-if" situations and testing different hypotheses. This modelling approach demonstrates great potential in demographic modelling, and we envisage that further work will show its utility across a wide variety of social science domains and policy applications.

References

Anderson, J. (1997). *Models for retirement policy analysis* (Report to the Society of Actuaries). Available at: http://www.soa.org/research/pension/research-models-for-retirement-policyanalysis.aspx. Accessed 10 Apr 2011.

Ballas, D., & Clarke, G. (2001). Modelling the local impacts of national social policies: A spatial microsimulation approach. *Environment and Planning C: Government and Policy, 19*, 587–606.

Ballas, D., Rossiter, D., Thomas, B., Clarke, G., & Dorling, D. (2005). *Geography matters: Simulating the local impacts of national social policies*. York: The Joseph Rowntree Foundation. ISBN 1 85935 266 9.

Birkin, M., Clarke, G., & Clarke, M. (1996). Urban and regional modelling at the microscale. In G. P. Clarke (Ed.), *Microsimulation for urban and regional policy analysis* (pp. 10–27). London: Pion.

Birkin, M., Turner, A., & Wu, B. (2006, June). *A synthetic demographic model of the UK population: Methods, progress and problems*. Proceeding of paper presented at The Second International Conference on e-Social Science, Manchester, UK.

Birkin, M., Procter, R., Allan, R., Bechhofer, S., Buchan, I., Goble, C., Hudson-Smith, A., Lambert, P., de Roure, D., & Sinnott, R. (2010). The elements of a computational infrastructure for social simulation. *Philosophical Transactions of the Royal Society A, 368*(1925), 3797–3812.

Bourguignon, F., & Spadaro, A. (2006). *Microsimulation as a tool for evaluating redistribution policies* (Working Paper 2006–20). Society for the Study of Economic Inequality. Available from http://www.ecineq.org/milano/WP/ECINEQ2006–20.pdf. Accessed 8 Apr 2011.

Brown, L., & Harding A. (2002). Social modelling and public policy: Application of microsimulation modelling in Australia. *Journal of Artificial Societies and Social Simulation, 5*(4). Available at http://jasss.soc.surrey.ac.uk/5/4/6.html. Accessed 22 Aug 2011.

Chin, S. F., Harding, A., Lloyd, R., McNamara, J., Phillips, B., & Vu, Q. (2005). Spatial microsimulation using synthetic small area estimates of income, tax and social security benefits. *Australasian Journal of Regional Studies, 11*(3), 303–336.

De Man, W. (1988). Establishing a geographic information system in relation to its use: A process of strategic choices. *International Journal of Geographical Information Systems, 2*(3), 245–261.

Gilbert, N., & Troitzsch, K. G. (2005). *Simulation for the social scientist*. Berkshire: Open University Press.
Hägerstrand, T. (1985). *Time-geography: Focus on the corporeality of man, society, and environment, the science and praxis of complexity*. Tokyo: The United Nations University.
Harding, A. (2007, 21 August). *Challenges and Opportunities of dynamic microsimulation modelling* Plenary paper presented to the 1st General Conference of the International Microsimulation Association, Vienna. Available at http://www.euro.centre.org/ima2007/programme/day2.htm. Accessed 1 July 2012
O'Donoghue, C. (2001). Dynamic microsimulation: A methodological survey. *Brazilian Electronic Journal of Economics, 4*. Available from: https://www.vengroup.com/ve-net/Library.nsf/ab283684d03f231d80256b520047d321/426E6F6D3E95DF4880256F6A0044E430/$file/Dynamic%20Microsimulation.%20A%20Methodological%20Survey.pdf. Accessed 5 Mar 2009.
ONS (Office of National Statistics, UK). (2001). Table VS4: Vital statistics for wards 2001–2001 boundaries, Produced by Program Annual_ICD10_VS4_2001.
Orcutt, G. H. (1957). A new type of socio-economic system. *Review of Economics and Statistics, 39*, 116–123.
Rees, P., Stillwell, J., & Tyler-Jones, A. (2004). The city is the people: Demographic structure and dynamics (Chapter 2). In R. Unsworth & J. Stillwell (Eds.), *Twenty-first century Leeds: Geographies of a regional city* (pp. 26–48). Leeds: Leeds University Press. ISBN 0 85316 242 5.
Rephann, T. (2001). *Economic-demographic effects of immigration: Results from a dynamic, spatial microsimulation model*. The 2001 Annual Meeting of the Mid-Atlantic Division of the Association of American Geographers, College Park, MD. Available at: http://www.equotient.net/papers/immmic.pdf. Accessed 10 Apr 2011.
Rowland, D. (2003). *Demographic methods and concepts*. New York: Oxford University Press. ISBN 019875263.
Tanton, R., Vidyattama, Y., McNamara, J., Vu, Q. N., & Harding, A. (2009). Old, single and poor: Using microsimulation and microdata to analyse poverty and the impact of policy change among older Australians. *Economic Papers: A Journal of Applied Economics and Policy, 28*(2), 102–120.
Van Imhoff, E., & Post, W. (1998). Microsimulation methods for population projection. *Population: An English Selection, 10*, 97–138.
Vidyattama, Y., Cassells, R., Harding, A., & Mcnamara, J. (2011). Rich or poor in retirement? A small area analysis of Australian private superannuation savings in 2006 using spatial microsimulation. *Regional Studies*, doi: 10.1080/00343404.2011.589829. Available online: http://www.tandfonline.com/doi/abs/10.1080/00343404.2011.589829
Williamson, P., Birkin, M., & Rees, P. (1998). The estimation of population microdata using data from small area statistics and samples of anonymised records. *Environment and Planning A, 30*, 785–816.
Wu, B. M., & Hine, J. P. (2003). A PTAL approach to measuring changes in bus service accessibility. *Journal of Transport Policy, 10*(4), 307–320.
Wu, B. M., Birkin, M. H., & Rees, P. H. (2008). A spatial microsimulation model with student agents. *Computers, Environment and Urban Systems, 32*, 440–453.

Chapter 12
Design Principles for Micro Models

Einar Holm and Kalle Mäkilä

12.1 Background

When looking back on about 30 years of model design for dynamic microsimulation, spatial microsimulation, and agent-based microsimulation, it is obvious that the software platform is not stable. The dream of using generic modeling packages, or at least generic software packages, instead of programming each application model from the ground up has not yet materialized as a generally available alternative, despite the calls from practitioners for greater cooperation in the construction of such expensive models (e.g., Harding 2007). For special classes of applications and for models with not too many object instances, promising efforts have emerged like the ModGen toolset from Statistics Canada (Statistics Canada, no date) and all the R-based simulation tools contained in the UrbanSim open-source project (Waddell and Ulfarsson 2004), just to mention two examples out of several development efforts. Most current applied dynamic microsimulation models so far, however, like the MOSART model in Norway (Fredriksen et al. 2011), the Australian APPSIM model (Bacon and Pennec 2007; Harding et al. 2010), and the Swedish SESIM model (Brouwers et al. 2011) are hard-coded for a specific national dataset and application range, as were the original Dynasim and .CORSIM models (Caldwell and Morrison 2000). Several observers have highlighted the problems when it comes to effective sharing of knowledge and development efforts, one late effort being the creation of EURODYN, the European network for dynamic microsimulation (Dekkers and Zaidi 2011).

Of course, this problem has caused an enormous amount of extra work for those who develop microsimulation models. Many quite useful and theoretically interesting models do not survive because of the development effort required. If the microsimulation

E. Holm (✉) • K. Mäkilä
Department of Social and Economic Geography, Umea University, 901 87 Umea, Sweden
e-mail: einar.holm@geography.umu.se; k.makila@bredband2.com

R. Tanton and K.L. Edwards (eds.), *Spatial Microsimulation: A Reference Guide for Users*, Understanding Population Trends and Processes 6, DOI 10.1007/978-94-007-4623-7_12,

community is going to continue and thrive, then we have to invent some clever methods to save mental resources.

The purpose of this chapter is to respond to this need by promoting design principles that have gradually emerged from our long-standing efforts to create agent-based dynamic microsimulation models in Sweden. In order to enable a judgment on the relevance of our specific experiences, a short story of our different models is presented below.

Based on Torsten Hägerstrands time geography (Hägerstrand 1970, 1991), one early model containing much of the intellectual content of later models was the "HÖMSKE" model developed as an application demonstrating an extension of time geography into a time geographic theory of action (Holm et al. 1989). That model contained individuals in families giving birth, dying, moving, getting educated, working, etc. The model contained 1,000 synthetically created individuals, the practical maximum for the computers of those times (a DEC-10 mainframe).

The next step was to access an early version our huge longitudinal individual database (ASTRID), now containing several generations of all Swedes with more than a hundred annually updated socioeconomic attributes, including place of living and working with 100 m of spatial resolution. This meant that there was no longer a need for a synthetic population to start a model with, and we could now estimate behavioral equations on a complete set of longitudinal individual data including individualized subsets of attributes of other persons and workplaces surrounding each person (e.g., the number of people with the same profession within a 50-km radius). The quest to model all Swedes individually soon emerged, in order to grasp the potential of the database. The network of relatives (based on mother and father pointers) is useless in a sample. A 10% random sample does not contain the mothers of 90% of the children. In addition, the distribution of many attributes and resources in the context of the sampled population rapidly becomes distorted if the information for an individual does not contain all object instances – however rare their attributes are.

However, this huge dataset and potential for modeling the whole population also created significant problems. In particular, the modeling needed to provide a representation of all individuals in the population on a not too specialized workstation. This effort resulted in a series of models jointly named the "SVERIGE" models (Holm et al. 1996, 2002, 2007; Clarke and Holm 1987; Holm and Sanders 2007). Specific applications include assessments of local and regional effects of nuclear waste disposal (Berner et al. 2011; Holm et al. 2008), plant shutdown (Rephann et al. 2005), immigration (Rephann and Holm 2004), diffusion of sick leave (Holm and Öberg 2004; Holm et al. 2004a), modeling pandemics (Holm and Timpka 2007), labor supply (Holm et al. 2004b), and the aging population (Strömgren and Holm 2004).

Dynamic microsimulation models invariably contain localized individuals and often interaction between individuals within and outside the family, and interaction between individuals and firms and schools, similar to the ambition of agent-based simulation as reported in Boman and Holm (2004) and Holm and Sanders (2007). A few models applied in countries outside Sweden further the techniques for creating synthetic populations from aggregate data and surveys while otherwise conforming to the SVERIGE tradition (Strömgren and Holm 2010; Aschan-Leygonie et al. 1999).

Other applications include the development of a "twin"-based model applied to exploring the development of the balance between local supply and demand for elderly care, a regional population projection system (ACPOP), and a model for assessment of the endogenous growth impact of infrastructure investments ("InfraSim").

12.2 The Wish List

There are a number of aims that we want to achieve when developing a spatial microsimulation model. These include:

- Using the most modern software
- Using standard methods, shared by many users
- Backward compatibility (so keeping our old models and subsystems running)
- Avoiding relearning
- Developing solutions that are theoretically well designed
- Transferring knowledge and know-how to new colleagues

Achieving all these aims fully in one model is going to be very difficult, as some of them could be taken as contradictory. For example, we want to use the most modern software, but we also want backward compatibility. When looking at the whole model, we can see three levels of components involved in the model building process:

1. The internal logic of the model (the kernel)
2. The input-output system and other general tools used in the model (the tools)
3. The surface of the model (the user interface and presentation of results)

At the extreme ends, the choice is simple: if you have found a useful and stable programming environment, keep using it. This advice applies at least for the kernel but also for the tools. While new and useful methods have been emerging using other environments, the cost of moving to a new environment can be significant in terms of programmer training in the new environment. Most end users also want to use the interface design of the current generation of software, particularly if there is menu-driven interaction with the program.

Most of the remarks below are connected to the tool level. We argue that the kernel and the tools should be kept apart but still kept in the same stable programming environment. These two together, which we refer to as the *application program*, should be strictly independent of the surface level – the user interface. That is not to say that the user interface has to remain the same for different application programs – typically, it is as easy to construct a tailor-made user interface in a certain application as it is to do it in Excel.

To design a generic package that enables model building without programming is desirable but so far unrealistic. It might be possible, but the risk is that it would end up as a new language that is far more technically complicated than the one that was being avoided. Instead, the model design can be constructed at a number of levels:

- Identify the fundamental functionality needed.
- Separate clearly the kernel of the model from the rest of the model, which will be either parameters for controlling the model experiments, code for handling the simulated objects, or output which enables the user to analyze the outcome of the experiments.
- Only elaborate the kernel and the user interface in detail as parts of the simulation program. These are the parts that are most specific to each application.
- The rest of the model (the tools) should be able to be shared between many models.

12.3 Quasi-Independent Subsystems

The choice of design principles and other functionality will be discussed under the following headings:

- Parameter input
- Matrix input
- Equation evaluator
- Result aggregation
- Biography aggregation
- Memory allocation
- Random number generation and use
- Handling of sets
- Random choice between many alternatives
- Primary and secondary attributes
- Parallel execution
- Twins or equations

In the models developed by us over the last 20 years, all of these subsystems have been part of the process. In a few cases, something that is close to a generic package has been emerging, i.e., a module that can be utilized as a ready-made component in later models. More typical has been the copying and modifying of code from one model to another, sometimes simplifying it, sometimes adding features required for the new application. Therefore, we have not presented these subsystems as existing packages or modules. Instead, we are content to contribute some methodological remarks concerning each of the points above.

12.3.1 Parameter Input

Parameters in this case signifies single values, either describing the world simulated, the experiment done, or references (e.g., file names) to more elaborate data. We recommend that this is kept very simple: just a text file with pairs (parameter name:parameter value). Of course, a more elaborate user interface is an advantage, but we prefer to keep this outside the application, and then it can take many different forms in the interface, so values can be provided by menu choices to support updates, and error checks can be used to avoid illegal values.

All of this can be designed very ambitiously if a model is used heavily and less ambitiously when it is used once or twice and then forgotten, but it is not part of the application. In order to get started with the most difficult parts of the modeling, it is better to start out with very simple parameter handling that can be copied from earlier models and then modified.

12.3.1.1 Matrix Input

Often, there are some fundamental data that occur in many models, typically two-dimensional tables (e.g., age and sex), or just one column of numbers or names. Input of these matrices can often be done by a standard procedure common to many models.

12.3.2 Equation Evaluator

It is convenient and relatively straightforward to implement behavioral equations so that they can be changed without recoding instead of being directly hard-coded, computing an expression like the following:

$$V = p_0 + p_1 * v_1 + p_2 * v_2 + \ldots p_i * v_i$$

where the p_i are estimated parameters and the v_i are values. Some of them are attributes of the simulated objects or other values dynamically calculated in the model. The final calculation is nearly always simple and can often be done by calling a general function. The parameters are then inputted from an external table, which is simple to do. The values are a bit more difficult: sometimes they are simply attributes of the simulated object and sometimes results of rather elaborate calculations based on these attributes and other state variables in the model. Most of this can be defined in a common top-level case statement where each case corresponds to one of the variables. Usually, we call this procedure *eval* and there is one eval function for each kind of simulated object.

12.3.3 Result Aggregation

There are two fundamental output streams from a microsimulation model:

1. The yearly object (population, firms, regions, etc.) status
2. The stream of events

These streams of events must be monitored through a set of well-designed functions that are called at the end of the year and whenever an event (a change of an attribute for a person or family) occurs. These procedures should produce aggregate information of a general multidimensional kind and also micro-data in a compact form. Any other ad hoc printing of output should be avoided. Ideally, the application will construct a set of standard tables from the aggregate data and print them on a set of standard files in a form that makes them directly suitable for post-processing and analysis work after the simulation. It is usually not sufficient to restrict this output to the most basic kind of log files at the microlevel. This will only add additional workload and delay for the user.

12.3.4 Biography Aggregation

A useful tool for debugging is biographies. A biography is all the historical data associated with a person in a microsimulation model. To be informative enough, biographies must be able to track a substantial part of the context of each event (e.g., for a birth, not only the id of the mother but also other attributes, also information on the father, the region of birth, etc.). The log files for biographical information will be very large and will have to be in a compact binary form (if not constrained to a chosen subset of the object instances), but still contain a lot of information to capture the context of each event.

12.3.5 Memory Allocation

For smaller samples (a few hundred thousand persons), it is possible to store all persons as individual objects in an object-oriented program like C++ and keep them in primary storage.

It is also possible to keep all the families in primary storage as class objects linked together in a linked list. Within each family, the members are also built as class objects and linked together in a list. This imposes an enormous overhead, considering that the memory allocation mechanism in C and C++ will use about 15 bytes extra for each item stored. There is also a large overhead in computing time to maintain these structures. However, it is not necessary to allocate all objects by individual calls to the C++ new function. Instead, a large buffer can be allocated by one call in the program, and then the persons in the sample should be allocated as objects through pointers into this buffer. This will save a lot of storage and also processing time.

When the number of objects reaches the level of millions, then another approach must be taken. They can still be kept in primary storage, but now they must be compacted as well. The internal program logic will then be similar to an old-fashioned data processing program where one family at a time is read from a file, and another file with the updated families is written. When reading the family, it is expanded to the normal problem-oriented class objects. And at the end of the year, the output file is closed and reopened as the input file for the next year.

So the processing is similar to reading and writing compressed or encrypted files, unpacking them one family at a time. However, the files are kept in primary storage, and there is one important reason for this apart from the processing speed for reading and writing on disk as compared to moving data from buffers. The reason is that it must sometimes (e.g., for the matching algorithms) be possible to reach any other person in the population while treating another person. This is solved by saving pointers to the starting points of the family and the relative number within the family for the person.

In the simplest kind of model, there is only one kind of person object. But within the engine that drives the scanning of the buffers, reducing the number of types of people can be used to save space. As an example, if we define people of three kinds: children, active adults, and passive pensioners. Only active adults will need a complete storage of all attributes. For the others, it is enough to set a flag that will cause a number of standard values to be assigned to certain attributes. Stored in this way, nearly 8.6 million persons – with about 50 attributes each – can be fitted into a buffer of about 340 Mb.

While constructing another model (InfraSim) with only about 15 attributes per person, an explicit comparison was made between directly using a conventional object representation of each of the nine million persons in Sweden and using a vector with 15 million (to cover expected population increase) elements indexed by person id for each of the 15 attributes. The object representation required close to 20 Gb of core memory. The vector representation fitted within 300 Mb, showing that significant increase in efficiencies can be gained using a vector representation.

12.3.6 Random Number Generation and Use

Very simple and efficient algorithms can be used to produce pseudorandom numbers. This section recommends some criteria for evaluating these algorithms:

- *Degree of distribution*. Usually, if one million numbers between zero and one are drawn, they will have a continuous uniform distribution. However, intervals near 0 and 1 may need to be checked closely to ensure a continuous uniform distribution.
- Check for autocorrelation between one number drawn and the next few. Ideally, this correlation should be very low, but there are algorithms where this correlation is high. Such algorithms may still work fine in many models as long as the continuous uniform distribution is maintained, but in some cases, they might cause very biased results. One example is when very small numbers tend to be followed by other small numbers. This might result in events with low probabilities

occurring more frequently than expected because they occur concurrently due to the autocorrelation.

- In many cases, these algorithms repeat an identical sequence after a specific number of draws. This is OK, but the number of draws used should not be too short.
- The inbuilt random number generator of some development environments sometimes deteriorates if used too many times with the same start seed. One remedy is to create a new random variable at certain points in time and execution.
- In some cases, simulations can be biased due to the order that the objects are processed in. Usually, the objects are stored in a specific order, and this order is kept from 1 year to the next. Often this is quite OK, but when the model involves matching (searching and forming combinations with other objects), the order may cause a bias. There is one simple remedy: scramble the whole database at the end of each year. This may sometimes take more time than the simulation itself, so make this optional so that scrambling is skipped during test and development.

12.3.7 Handling of Sets

There are many general tools for this in the Microsoft class libraries. Sometimes they are OK, but when dealing with large sets of objects, care needs to be taken. In many cases, it is better to design some code yourself. When doing this, efficiency with storage and computing time is essential. Bitmaps are useful for representing sets of objects, provided they have id numbers as dense series of integers. Arrays are usually efficient and it is not too difficult to design a class where arrays of dynamic size are allocated and expanded when needed. Care needs to be taken with structures involving lists. They are easy to create but might be wasteful of space. In particular, removing elements might be a waste of time. For each deleted item in the list, the method wrapper copies all elements but the removed one into a new vector and then replaces the old vector with the new one. To do that a million times within a loop through, the whole population is an enormous waste of computing resources. It is possible to keep using lists if the remove method is avoided and replaced by giving the elements to be removed a flag and then performing the actual remove by copying as if the list was a vector, i.e., at the end of the year.

12.3.8 Random Choice Between Many Alternatives

Sometimes in a model, choices are made between many alternatives according to a probability distribution. If this is done only a few times, then it can be done in a loop where a drawn random number is compared to the limits in a cumulative distribution. But when this is done for a large population and the number of choices is large (like when choosing between the 290 local administrative areas of Sweden, the municipalities), then this will take a lot of time. The process can be sped up considerably by generating an auxiliary data structure before the simulation is started or at the start of each year if the probabilities change over time. To do this, the program needs to generate a large vector of values, and for each alternative, store it repeatedly in proportion to the probability for that alternative. Then a choice can be made by just drawing one single random number. In the case of migration between municipalities, the large municipalities will have a large number of alternatives, so then a large vector is needed to represent them all with some precision. Smaller municipalities will have fewer alternatives so then a smaller vector can be used. While these vectors will take a lot of memory space, the gain in computing time will be enormous.

The same method can be used for other attributes, e.g., professions, and also in matching procedures, combining persons with other persons or with workplaces.

As a general point, there has been a shift of balance between storage and computing time. We are often impressed by the great improvement in computing time but tend to forget that the improvement in terms of memory in computers, both in terms of size and price, is even bigger. A large amount of memory can now be devoted to auxiliary data that will speed up the search in large sets of data.

12.3.9 Primary and Secondary Attributes

The simplest models only deal with one kind of object (the person). Other models contain several object classes like family, workplace, and municipality. The additional objects may possess static as well as dynamic attributes which can change during the simulation (e.g., the number of members in families or workplaces, or the mean income in a workplace). Persons have their own attributes, but are also connected by pointers to other objects, and so they can access attributes of the other objects. All of these attributes can be used in the model equations. This means all attributes should be accessed in the same way through a function (eval) containing a case statement with cases for both the primary attributes and other cases that will call the eval functions of the connected objects or other functions that make additional calculations to transform or recombine the stored attributes. The inbuilt "property" construct might sometimes give a useful alternative.

12.3.10 Parallel Execution

One obvious emerging option to increase the speed of a large simulation is to make use of the threading capabilities of current multi-core computers. There are now tools available to organize this at the source code level, e.g., by partitioning the program and assigning different processors to simulate these partitions in parallel. This works only if each partition is strictly independent of the others, so it's easy to achieve only for the simplest kinds of models. We have moved towards more and more interaction between different parts of the model, so these mechanisms are not easy to use in our models without complicating the source code considerably.

For certain parts of the program, high-level methods like the canned parallel for loop of the .NET framework are sometimes useful. One advantage of using such constructs is that it just doesn't work if the parameters of the called function are not strictly independent of the calculations in the function. That gives a kind of additional run time consistency check.

The increased speed seen using parallel execution is larger for the development phase than it is for final production runs of an application. During development, one very soon approaches the situation when it is necessary to run the full model many times with the entire population in order to discover remaining bugs. This can be very tedious if the compile–run cycle takes a long time.

A typical production run often requires many replications and/or different alternative experiments. In these cases, it is easier to use the parallel machinery at process (operator) level. The simplest solution is to run several simulation experiments in parallel. Another way that was implemented in the SVERIGE model is to implement some of the housekeeping activities needed to collect aggregates during the simulation or to maintain auxiliary data that supports rapid access in memory as parallel processes running concurrently with the simulation.

12.3.11 Twins or Equations

A demanding analytical task is to accurately model the outcomes for large choice sets, like the choice of destination when moving geographically or the choice of specific education or profession or place of work. Geographic mobility and connected family changes are core events in spatial population modeling. A Wilson/Fotheringham-type interaction model tends to require fine-grained alignment in order to produce reasonably accurate local outcomes, almost to the point where the output largely reproduces the alignment factors. Therefore, especially in situations with a large and diverse choice set like destination choice combined with simultaneous family changes, imitating empirical twins might stand out as a simple viable alternative to multiple logit-based equations or interaction models. Instead, a random assignment scheme is applied based on imitating observed behavior of similar individuals ("twins") as discussed by Klevmarken (1997).

We have tested this alternative as demonstrated by the following simple example model tested on a database of the Swedish population. The example model contains three time-independent variables (birth year, sex, and place of birth) and four time-dependent variables (municipality, education level, civil state, and disposable income). Municipality is fully represented (290 values) but the other variables have been reduced to make a complete match of twins possible. Twins are picked for one particular year in the example (1993). The year after the twin year is the result year. So the whole simulation procedure can be described as:

1. Scan the whole population and locate a twin with an identical or very nearly identical set of attributes.
2. Pick up the set of attributes for this twin in the result year.

3. Let the simulated person inherit the whole set of attributes as values for the next simulated year.

Altogether, there are 2*4*512*290 = 1,187,840 possible combinations for the time-dependent variables distributed over 2*4*64 = 512 combinations of the time-independent variables (sex, origin, and age class). Before the simulation can be done, the behavior of the individuals for the past years must be aggregated in such a way that those having a specific combination can be easily accessed. For each combination occurring in the dataset, a number of people are found, each of them having a particular combination of result variables. All of this must be loaded in the core before the simulation.

Note that twins are not unique or limited to just a few individuals. Many of the technically possible 1,187,840*512 cells are actually empty, but others might contain thousands of individuals. It is not always possible to find an exactly matching twin. In the example using data for the Swedish population, we applied a set of strategies for locating "proxy twins" which are those that are similar except for one or two dimensions. When simulating about nine million individuals, we get approximately the following outcome in terms of hits:

In full detail	7,000,000
Within age +/−1 or +/−2	1,000,000
Within proximate income	600,000
In the most frequent income	100,000

The hypothesis is that twin replication might sometimes outperform behavioral equations, especially when the outcome is complex, not binary or scalar like the destination choice of movers; when several events interact simultaneously like family formation and mobility; or when it is important to maintain a realistic heterogeneity in the long run like an income distribution.

Advantages of using twins are that heterogeneity in all attributes is maintained automatically and that latent information not obvious from the attributes can affect the results. In addition, twins might in some cases function as a consistent alternative to alignment.

The disadvantages of using twins' replication are that results easily get locked into the sample of the empirical twins. Problems can occur when the simulation approaches a state not experienced by any empirical twin. Experiments changing behavior are easier to perform using analytical behavioral equations. In addition, the definition of and the criteria for selecting a twin are not obvious except in very simple cases.

It has been demonstrated that for a simple simulation, twin replication produced somewhat superior results at least for the age distribution of movers and the destination of movers. For complex events, it might in some cases give a convenient alternative to equations that maintain consistent heterogeneity better than using a table look up. Questions like how to define similar, how to select twins, and how to use biographical individual information still need to be solved if using models with individuals with many attributes that make them unique. So, in this case, we have no general recommendation to make.

12.4 Conclusion

Our own answer to the relevance question raised in the background section is that most of the discussed design principles might be even more relevant in other countries which don't have the rich longitudinal individual data that exists in Sweden. However, these principles are recommendations about how to code and reuse code, and there are no ready to use software modules or modeling packages for dynamic spatial microsimulation. We would recommend that these principles be implemented while constructing new national and regional simulation models, and we would particularly emphasize the use of the proposed design principles in efforts to create generic high-level software for dynamic microsimulation.

One alternative would be to integrate core parts of the discussed design principles into something like the already well-developed toolset of ModGen. This would then contribute somewhat towards moving the resulting software into a tool, replacing the need for a large portion of our own (as well as others') application specific coding for different models. It would then be possible for the next generation of social scientists in academia and in agencies to apply micro-based dynamic modeling on urgent problems instead of entirely relying on regression analysis with its obvious shortcomings when it comes to representing the complex dynamic interactions of different agents within society.

References

Aschan-Leygonie, C., Baudet-Michel, S., Gautier, D., Holm, E., Lindgren, U., Mäkilä, K., Mathian, H., & Sanders, L. (1999). Micro modelling of the population dynamics in a region with strong urban growth. In S. E. Van der Leeuw (Ed.), *Archeomedes*.

Bacon, B., & Pennec, S. (2007). *APPSIM – Modelling family formation and dissolution* (Online Working Paper – WP2). http://www.canberra.edu.au/centres/natsem/

Berner, B., Drottz Sjöberg, B., & Holm, E. (2011). *Social science research 2004–2010, themes, results and reflections*. Stockholm: SKB. ISBN: 978-91-978702-2-1

Boman, M., & Holm, E. (2004). Multi-agent systems, time geography, and microsimulations. In M.-O. Olsson & G. Sjöstedt (Eds.), *Systems approaches and their application: Examples from Sweden* (pp. 95–118). Dordrecht: Kluwer.

Brouwers, L., Ekholm, A., Janlöv, N., Johansson, P., & Mossler, K. (2011, June 8–10). *Simulating the need for health- and elderly care in Sweden – A model description of Sesim-LEV*. Paper presented at the Third General Conference of the International Microsimulation Association Stockholm.

Caldwell, S., & Morrison, R. (2000). Validation of longitudinal microsimulation models: Experience with CORSIM and DYNACAN. In L. Mitton et al. (Eds.), *Microsimulation in the new millennium*. Cambridge: Cambridge University Press.

Clarke, M., & Holm, E. (1987). Micro-simulation methods in spatial analysis and planning. *Geografiska Annaler Series B, Human Geography, 69*(2), 145–164.

Dekkers, G., & Zaidi, A. (2011). The European network for dynamic microsimulation (EURODYM) – A vision and the state of affairs. *International Journal of Microsimulation, V4*(1), 100–105.

Fredriksen, D., Knudsen, P., & Martin Stølen, N. (2011, June 8–10). *The dynamic cross-sectional microsimulation model MOSART*. Paper presented at the Third General Conference of the International Microsimulation Association Stockholm.

Hägerstrand, T. (1970). *What about people in regional science, regional science association papers, Vol. XXIV*. Heidelberg: Springer.

Hägerstrand, T. (1991). Tiden och Tidsgeografin. In G. Carlestam & B. Sollbe (Eds.), *Om tidens vidd och tingens ordning, T21*. Stockholm: Statens råd för byggnadsforskning.

Harding, A. (2007, August 21). *Challenges and opportunities of dynamic microsimulation modelling*. Plenary paper presented to the 1st General Conference of the International Microsimulation Association, Vienna. Available at http://www.euro.centre.org/ima2007/programme/day2.htm

Harding, A., Keegan, M., & Kelly, S. (2010). Validating a dynamic microsimulation model: Recent experience in Australia. *International Journal of Microsimulation, 3*(2), 46–64.

Holm, E., & Öberg, S. (2004). Contagious social practice? *Geografiska Annaler, 86B*(4).

Holm, E., & Sanders, L. (2007). Spatial microsimulation models. In L. Sanders (Ed.), *Models in spatial analysis* (Geographical information systems series). Newport Beach: ISTE.

Holm, E., & Timpka, T. (2007). A discrete time-space geography for epidemiology: From mixing groups to pockets of local order in pandemic simulations. *Studies in health technology and informatics, 129*, 464–8.

Holm, E., Mäkilä, K., & Öberg, S. (1989). *Tidsgeografisk handlingsteori – Att bilda betingade biografier*. Gerum Rapport 8, Department of Geography, Ume University.

Holm, E., Lindgren, U., Mäkilä, K., & Malmberg, G. (1996). Simulating an entire nation. In G. Clarke (Ed.), *Microsimulation for urban and regional policy analysis*. London: Pion.

Holm, E., Holme, K., Mäkilä, K., Mattsson-Kaupi, M., & Mörtvik, G. (2002). *The SVERIGE spatial microsimulation model – Content, validation, and example applications* (Gerum Kulturgeografi 2002, Vol. 4). Umeå: Umeå Universitet.

Holm, E., Lindgren, U., Eriksson, M., Eriksson, R., Häggström Lundevaller, E., Holme, K., & Strömgren, M. (2004a). *Transfereringar och arbete* (Arbetsrapport R2004, Vol. 16). Östersund: ITPS – Institutet för tillväxtpolitiska studier.

Holm, E., Lindgren, U., & Malmberg, G. (2004b). *Arbete och tillväxt i hela landet – betydelsen av arbetskraftsmobilisering* (Vol. 22). Östersund: ITPS – Institutet för tillväxtpolitiska studier.

Holm, E., Lindgren, U., Häggström Lundevaller, E., & Strömgren, M. (2007). SVERIGE. In A. Gupta & A. Harding (Eds.), *Modelling our future, population ageing health and aged care: International symposia in economic theory and econometrics* (Vol. 16). Amsterdam/Boston: Elsevier.

Holm, E., Lindgren, U., & Strömgren, M. (2008). Socioekonomiska effekter av stora investeringar i Oskarshamn. SKB R-08–76.

Klevmarken, A. (1997). Behavioral modeling in micro simulation models: A survey (Working Paper Series 997:31). Uppsala University, Department of Economics.

Rephann, T., & Holm, E. (2004). Economic-demographic effects of immigration: Results from a dynamic, spatial microsimulation model. *International Regional Science Review, 27*, 379–410.

Rephann, T., Mäkilä, K., & Holm, E. (2005). Microsimulation for local impact analysis: An application to plant shutdown. *Journal of Regional Science, 45*, 183–222.

Statistics Canada. (no date). *Modgen 15 years of creating models*. http://www.statcan.gc.ca/microsimulation/pdf/modgen-hist-eng.pdf. Accessed 20 Aug 2011.

Strömgren, M., & Holm, E. (2004). *Åldrande befolkning och framtida behov av kommunalskatt*, Kulturgeografiska institutionen, Umeå universitet.

Strömgren, M., & Holm, E. (2010). *Using downscaled population in local data generation* (Technical Report). ESPON 2013 Database.

Waddell, F., & Ulfarsson, G. (2004). Introduction to urban simulation: Design and development of operational models. In B. Stopher & H. Kingsley (Eds.), *Handbook in transport, Volume 5: Transport geography and spatial systems* (pp. 203–236). New York: Pergamon Press.

Chapter 13
SimEducation: A Dynamic Spatial Microsimulation Model for Understanding Educational Inequalities

Dimitris Kavroudakis, Dimitris Ballas, and Mark Birkin

13.1 Introduction and Background

This chapter presents a dynamic spatial microsimulation approach to the analysis of educational inequalities. The method simulates individual units over a period of time creating a picture of the evolving attributes of each unit. As Birkin et al. (1996) point out, the updating of microsimulation populations would typically involve list processing based on either deterministic rules (for instance, population ageing, or change in the allocation of family benefits) or a probabilistic change of states (e.g. what is the probability that an individual will get married next year, given his/her socio-economic and demographic situation). There are a series of demographic and socio-economic transitions that can be modelled with the use of spatial microsimulation. In other words, dynamic spatial microsimulation projects each microunit to a future state altering its attribute values according to predefined rules. These rules relate to the unit's population changes. For a population in any geographical area, those rules may be functions about population ageing or mortality and birth rate. These rules reshape the dataset over time (annually or monthly) and create a new estimated snapshot of the population at these time points. Dynamic spatial microsimulation models can be used to model transitions such as leaving home, entering school, university, the labour market, etc. As Gilbert and Troitzch (1999) point out:

D. Kavroudakis (✉)
Department of Geography, University of the Aegean, Mytilene, Greece
e-mail: dimitrisk@geo.aegean.gr

D. Ballas
Department of Geography, University of Sheffield, Sheffield, UK

M. Birkin
School of Geography, University of Leeds, Leeds, UK

R. Tanton and K.L. Edwards (eds.), *Spatial Microsimulation: A Reference Guide for Users*, Understanding Population Trends and Processes 6, DOI 10.1007/978-94-007-4623-7_13,

> During their lifetimes, the simulated individuals have to change their educational and employment status. They will enter school with different probabilities when they are between 14 and 20 years old, they will be employed in different jobs, lose their jobs, earn an income which depends on their type of job, and eventually retire with different probabilities depending on their ages. (Gilbert and Troitzsch 1999, p. 59)

For the initialisation of a spatial microsimulation process, a number of specialised datasets are required (see Chap. 2 for a general description of data required and data preparation). These datasets should include a detailed sample of the population. This sample is then used to create a microdataset which is the initial population for the microsimulation process. One of the major problems with the dynamic modelling of socio-economic events is the lack of relevant data on transition probabilities at appropriate geographical scales (e.g. the probability that an 18-year-old male of a given socio-economic background living in a particular locality would migrate; such data are typically available at a national or coarse geographical scale). One way of tackling this data paucity issue is by trying to formulate plausible assumptions about the socio-economic transition probabilities of different individuals and households and incorporate them into the model. Such assumptions could be formulated on the basis of the UK British Household Panel Survey (BHPS) which is a rich longitudinal population survey, with a large nationally representative sample, which has been conducted annually since 1991 (Taylor et al. 2001).

The work presented in this chapter builds on relevant microsimulation research (Ballas et al. 2005a, b; Rossiter et al. 2009) utilising the BHPS in order to build a dynamic spatial microsimulation model for the analysis of social and spatial inequalities in educational attainment. It can be argued that the subject of educational attainment is particularly suitable for the development and application of a dynamic spatial microsimulation model given the influence that education has on a person's life outcomes. Additionally, there are important policy implications pertaining to social and spatial inequalities that can be explored with such a model. In particular, there is increasing evidence (Wilkinson and Pickett 2009) that the reduction of socio-economic inequality should be a major social policy goal in modern societies. Educational attainment and a person's level and type of educational achievements typically have a major influence on where he/she will end up in the distribution of those potentially life-enhancing "goods" (Clarke 2003). In turn, access to educational opportunities is influenced by both socio-economic factors as well as by economic and social geography.

Educational outcomes at personal and community levels have been well researched (Hall 1961; Fryer 1997; Demack et al. 2000; Allen and van der Velden 2001). Subsequently, there is also a need to understand inequality geographically in order to be able to prevent income-related inequalities in educational opportunities. The use of spatial microsimulation models can be used to produce suitable output for a deeper understanding of inequalities and forms a valuable tool for the prediction of such dynamics.

Figure 13.1 highlights the importance of household income and geography with regard to educational opportunities and outcomes. It has long been argued that the social and spatial divisions and inequalities in educational attainment can be seen as

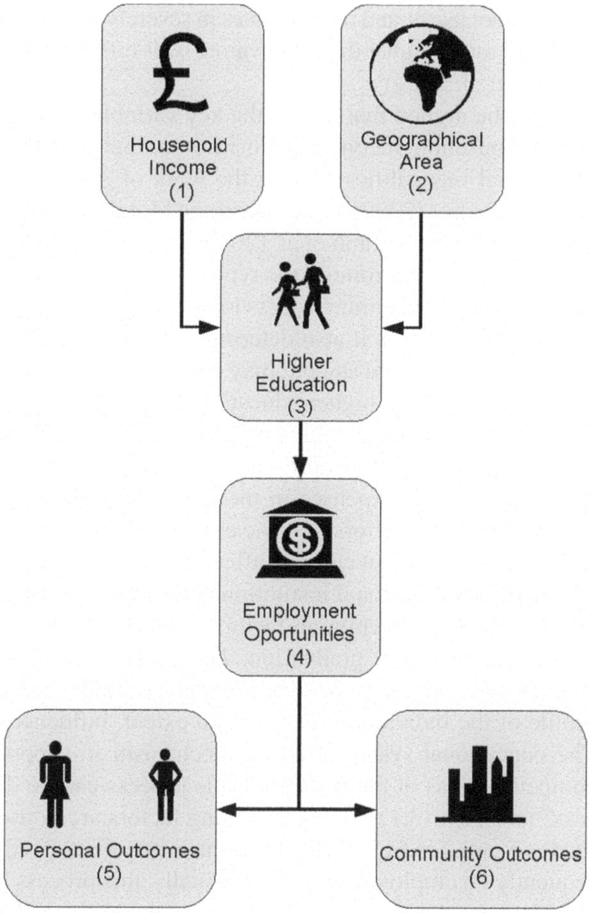

Fig. 13.1 Modelling educational attainment

the geographical manifestations of deeper social and spatial inequalities and divisions in relation to a wider range of indicators (Kasarda 1993; Pacione 1997; Ainsworth 2002; Thomas et al. 2009).

There are a number of empirical studies throughout the past two decades that paint a story of social and spatial divisions in educational opportunity and attainment. In particular, there has been considerable research on the impact of space and place upon educational attainment and in particular primary and secondary educational attainment with the use of statistical modelling techniques (see Demack et al. 2000; Ainsworth 2002; Clarke and Langley 1996; Gibbons 2002; Gibbons and Machin 2003; Harris and Johnston 2008). Most of this work focuses on the influences of neighbourhood, the environment in which a child is raised and upon primary and secondary educational opportunities and outcomes. In particular, it has long been argued that neighbourhood characteristics have a very strong impact on life chances through their influence on the educational outcomes of young residents (Ainsworth 2002).

Early school leaver and dropout rates in severely distressed or deprived neighbourhoods tend to be much higher than in the more well-off parts of a city or region (Kasarda 1993; Thomas et al. 2009).

It can be argued that one of the key variables associated with educational attainment is household income. There are a number of studies on income inequalities and educational inequalities such as the work of Azzoni and Servo (2002), Rodríguez-Pose and Tselios (2009), López Bazo and Motellón (2009), Duranton and Monastiriotis (2001, 2002) and Miranti et al. (2010). It can be argued that household income plays a major role in determining the type of higher education a person could have access to. As well as determining access to better schooling opportunities (e.g. see Cheshire and Sheppard 2004), it also determines the type and number of educational institutions that the potential student may apply to (e.g. see Singleton 2009). Another determining factor of the higher educational process is the geographical area a person is brought up and lives in, which influences educational availability by determining the available educational institutions a potential student may apply to due to proximity and/or living cost expenses in the area where the institution is located (Singleton 2009). These two factors to some extent affect the process of selecting a higher educational institution but can also affect the selection of the type and topic of study. The choice of an educational institution determines the value and robustness of a degree which will be a determining factor, to some extent, of the type of job that the individual will find after graduation. The skills of all economically active units of the economy can then in a way determine the profile of the economy. In other words, the profile of the individuals, to a certain extent, influences the profile of the economy. The educational system offers a mechanism to support the economic growth and competitiveness of an economy. This process can be described as a "waterfall process" (see Fig. 13.1). The determining factors are household income and geography (boxes 1 and 2 in Fig. 13.1). These have an effect on education (box 3) and this subsequently on employment (box 4). Finally, this process has spin-offs ("by-products") both for the society and the individuals themselves. These are the major milestone stages that are the subject matter of the research presented in this chapter.

The remainder of this chapter is organised as follows: the next section describes the data that were used to calculate the probabilities that drive the dynamic model (known as "transition probabilities"). This is followed by a third section describing the dynamic model in more detail. The fourth section presents and discusses some model outputs, and the final section offers some concluding comments.

13.2 Using Secondary Data to Estimate Transition Probabilities

The model presented here is the dynamic component of a spatial MSM approach, comprising both a static and a dynamic model (Kavroudakis 2009). The model used a combination of data from both the BHPS, which contains a wealth of socio-economic and demographic variables (Taylor et al. 2001) and small area statistics tables from the census of population (Census Dissemination 2008). These census data are the product of the most authoritative social accounting of people and housing in Britain of its time (Cole et al. 1993). However, the census records demographic and socio-economic information at a single point in time and is therefore less appropriate for the study of social and economic changes through time. Conversely, the BHPS is a very useful tool for the understanding and analysis of social and economic change at the individual and household level, as well as to identify, model and forecast such changes and their causes and consequences in relation to a range of socio-economic variables (Taylor et al. 2001). Of particular relevance here is the wealth of information in the BHPS about educational qualifications, which can be categorised by social class and sex and analysed annually from 1991 (when the BHPS was launched) onwards. The BHPS offers a valuable resource for understanding the underlying mechanisms which influence the life events that can be simulated like graduation from higher education, finding a job, marriage and mortality. Nevertheless, the BHPS dataset is disaggregated at a relatively coarse level of geography (the smallest area for which any meaningful analysis can be conducted for the whole country is the region). This is to preserve confidentiality as the questions answered in the survey are highly sensitive and contain private information that would not be able to be collected without a prior assurance of confidentiality. However, this level of geographic detail is not sufficient enough for a detailed spatial analysis.

A thorough statistical descriptive analysis of the data in the BHPS was undertaken in order to explore the interdependencies between socio-economic classification and other variables that could potentially be included in the spatial MSM. In building a dynamic spatial microsimulation model of educational attainment, it is important to first explore the temporal patterns in the relationships between key variables such as "higher academic qualification", socio-economic background and income. It should be noted that one of the major advantages of using the BHPS is that an individual can be tracked through all the waves. Waves are the annual datasets of British Household Panel Survey that contain all data collected for that specific year. Responders were selected randomly in the first wave (1991) and ranged from all ages, income groups and geographical locations in order to have a diversity of respondents and behaviour patterns (IISER 2006). The same individuals and households are then interviewed every year (or at least there is an attempt to do so) generating a wealth of longitudinal information.

The BHPS can thus be used to create educational life paths in order to track the individuals in every panel wave and record any change in their educational qualifications. The illustrated life paths are just an indication of the educational mobility of the individuals. Educational progress is also associated with other social characteristics which determine the choice and output of an educational process. The BHPS variables "Goldthorpe Class: most recent job" and "Highest Academic Qualification" from all BHPS waves were used to create such "educational life paths". These are the representation of educational qualifications with respect to the waves of BHPS data. Table 13.1 shows the different social class and academic qualification categories as well as the proportion of individuals falling within each category in the first wave of the BHPS.

Table 13.1 Social class and highest academic qualification categories (from Taylor et al. 2001)

Goldthorpe social class categories	Frequency (%)	Highest academic qualification categories	Frequency (%)
Service class, higher (SC1)	10.4	Higher degree	1.2
Service class, lower (SC2)	16.6	1st degree	5.8
Routine non-manual (SC3)	14	HND, HNC teaching	4.8
Personal service worker (SC4)	7.8	A level	13.3
Small proprietors with employees (SC5)	1.6	O level	24.5
Small proprietors without employee (SC6)	4.6	CSE	5.2
Farmers, smallholders (SC7)	0.7	None of these	41.7
Foreman, technicians (SC8)	7.1	Missing/proxy respondent	3.6
Skilled manual workers (SC9)	8.5		
Semi-skilled, unskilled, manual (SC10)	21.4		
Agricultural workers (SC11)	0.8		
Missing/proxy respondent	1.9		
Never had a job	4.6		

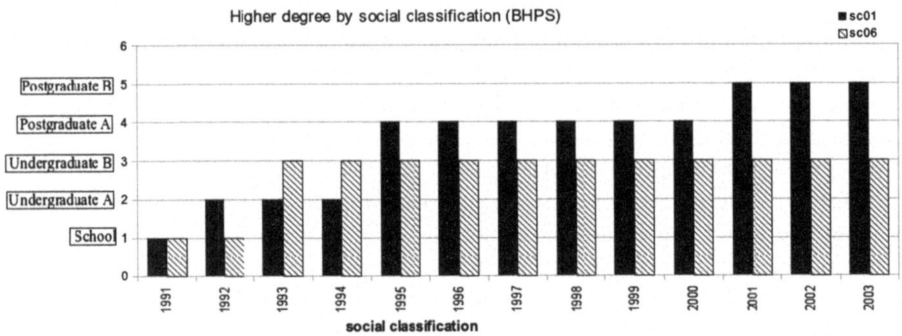

Fig. 13.2 Mean educational attainment of all individuals in social class 01, 1991 to 2003 (Source: BHPS Dataset)

All individuals from every social class were selected and tracked through the available data waves. Then average educational attainment for every social class was constructed using the highest educational qualification obtained for individuals in every social class every year from 1991 to 2003. As can be seen in Figs. 13.2 and 13.3, there is a pattern concerning the "educational mobility" of social classes (see Table 13.1 for more details on the social class categorisations). This pattern suggests that it is more likely for individuals from more affluent social classes to end up with a higher educational qualification compared to individuals from less-affluent backgrounds. This is the kind of information that can be utilised in a dynamic spatial microsimulation model to simulate events such as "going to university" and "graduating from university" at the microlevel. In particular, the BHPS is very suitable

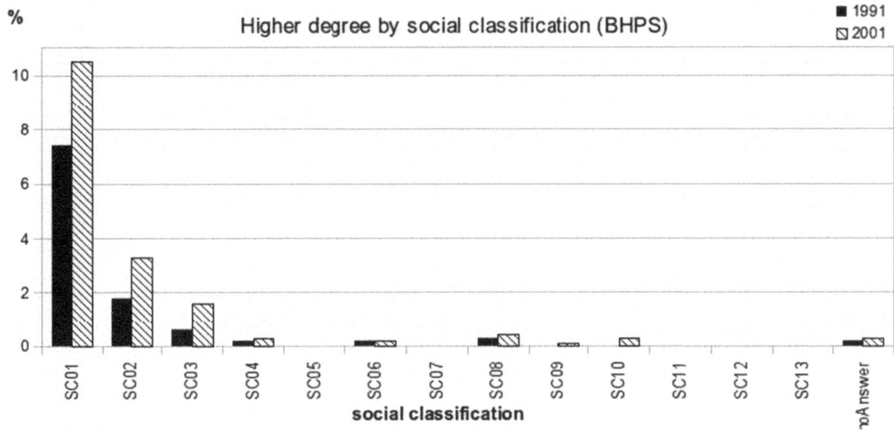

Fig. 13.3 Higher education qualifications by social class, 2003 (Source: BHPS dataset)

for dynamic microsimulation as it gives insights into what factors and variables are important with regard to individual and household life course dynamics. In addition, data such as that presented in Figs. 13.2 and 13.3 can be used to calculate the so-called transition probabilities that provide an estimate of how likely an individual is to undergo a transition from one state to another at one point in time. For example, we can calculate the probability of a 17-year-old individual from a "higher, service class" household entering university at age 18, graduating at age 21 and having a PhD degree at age 24. Such probabilities were calculated to perform the dynamic spatial microsimulation.

For instance, in order to evaluate educational life paths, the percentage of each social group in the BHPS with a higher degree was analysed. It was found that a larger percentage of social groups 1, 2 and 3 (more affluent) have a higher educational degree in both 1991 and 2003. This corroborates the argument that there is a positive correlation between household income and educational attainment.

Probabilities from the BHPS such as those described above were combined with probabilities by household income and were used in order to simulate university entry for every individual in our microsimulated database. In addition, data from UCAS have been used to determine the selection of the subject for each of the simulated individuals that enter higher education. Table 13.2 presents an extract of the information used by the model, showing the subject for individuals applying to universities in the Yorkshire and Humber region. The model utilised these data to assign a subject to each individual simulated to enter university.

We also calculated employment and unemployment probabilities by age, sex, educational qualification and social class using data from the BHPS. Similarly, we utilised data from the ONS to calculate transition probabilities for mortality. These probabilities can be used to formulate dynamic microsimulation rules and to then evaluate all individuals every year against those rules (following an approach similar to that of Ballas et al. 2005a). Overall, we calculated probabilities for the following transitions:

Table 13.2 Percentage of accepted applicant by subject area 2006 (UCAS 2008)

Subject	Postgraduate (%)	Undergraduate (%)	Total (%)
Medicine and dentistry	30.8	69.2	2.3
Medicine related	15.0	85.0	12.3
Biological sciences	18.7	81.3	6.1
Vet. science, agriculture and related	15.3	84.7	1.0
Physical sciences	24.5	75.5	3.3
Math and computing sciences	18.5	81.5	6.2
Engineering and technology	25.5	74.5	6.0
Architecture, building and Planning	23.7	76.3	2.5
Social sciences (inc law)	23.0	77.0	11.8
Business and administrative	27.8	72.2	15.5
Communication and documentation	17.9	82.1	2.2
Languages	12.5	87.5	5.6
History and philosophy studies	17.6	82.4	4.2
Creative arts and design	9.8	90.2	6.5
Education	50.5	49.5	8.3
Other subjects	1.8	98.2	6.1
Unknown	0.0	100.0	0.0
All subjects	21.7	78.3	100.0

- Entering university by sex, household social class and household income (using data from the BHPS)
- Assignment of subject of study (using data from UCAS)
- Graduating from university by household social class (using data from the BHPS)
- Mortality (using ONS vital statistics)
- Employment (using data from the BHPS)

The following sections discuss the dynamic model in more detail and describe some model outputs. It should be noted that the modelling exercise presented in this chapter focused on the region of Yorkshire and the Humber which contains several universities and higher education colleges, including the Universities of Leeds, Sheffield and York that make up the White Rose University Consortium (2010) that funded the research presented here.

13.3 Dynamic Model

The dynamic spatial microsimulation model is based on rules used for the annual transitions of simulated individuals as described in the previous section. In every year of the simulation, the population of the geographical area is filtered by applying some rules. Those rules are the transition rules which determine the transition probabilities for each individual. As discussed in the introduction, household income is a key variable affecting higher education attainment. Therefore, it can be argued

Table 13.3 Household categorization on the basis of median income

Very poor	$Y \leq \frac{1}{2}m$
Poor	$\frac{1}{2}m < Y \leq \frac{3}{4}m$
Below average	$\frac{3}{4}m < Y \leq m$
Above average	$m < Y \leq m + \frac{1}{4}m$
Affluent	$m + \frac{1}{4}m < Y$

(Y = household income, m = median household income in the Yorkshire and the Humber region; income data based on the BHPS variable wFIHHYR "annual household income"; for more details on all BHPS variables, see Taylor et al. (2001))

that this is a key variable that needs to be estimated at the small area level using spatial microsimulation and to then be used as a factor in subsequent calculations of transition probabilities. Once estimated, household income can then be used to classify households between different income categories. In the context of the research presented here, we adopted a method of categorising microsimulated incomes using the median of the samples to split them into five groups, as shown in Table 13.3 (as presented by Ballas 2004 and also used by Ballas et al. 2007).

We then calculated education-related transition probabilities by income and social class. In particular, as noted at the end of the previous section, we calculated probabilities from secondary data to model the following events: university entry, assignment of subject study, graduation, mortality and employment/unemployment. Income was a key variable with regard to entering university and graduation, whereas holding a degree was a key variable determining the transition to "employment". For each event a Monte Carlo sampling process was adopted, meaning that a computer-generated random number was compared to the calculated transition probability for every simulated individual. If the number was smaller than the probability, then the individual would be assigned the respective event (for more details on how such a process is implemented, see Ballas et al. 2005a). The following rules were implemented with regard to each event:

Higher education entry: This rule determines how many individuals will enter higher education every year of the simulation. The model selects all potential students from each area (individuals with A levels and aged 16–19). Then according to the transition probabilities collected from the BHPS dataset by income group and social class, the model determines whether the individual will enter higher education.

Graduation: Again, social class and household income from the BHPS were used to calculate how many individuals will graduate each year from the educational institutions in the geographical area.

Table 13.4 Results from the microsimulation model: percentage joining higher education, graduating and finding a job by income group

	Very poor (%)	Poor (%)	Below average (%)	Above average (%)	Affluent (%)	*Total (%)*
Initial model population	20	19	21	20	20	*100*
Join HE	16	17	19	19	20	*91*
Graduation	15	15	18	18	20	*86*
Employment	15	14	18	17	18	*82*

Mortality: Vital statistics on mortality by age and sex were used to evaluate each simulated individual for mortality, following a similar procedure to that presented by Ballas et al. (2005a).

Employment: Every simulated individual was evaluated for moving to employment based on a Monte Carlo sampling process using transition probabilities calculated from the BHPS.

The dynamic spatial MSM projects the population to a future state by applying annual transition probabilities to the population. A summary of some of the results of that process is depicted in Table 13.4 which shows the percentage of each income group from the initial simulated population that joins higher education, graduates and then finds a job. The initial model state has almost 20 % from each income group. The next step ("join HE") uses the results from the spatial microsimulation model to determine the number of individuals that hypothetically enter a higher education institution.

It should be noted that the dynamic spatial microsimulation model presented here is relatively basic and at an experimental stage. There is a need for thorough validation of the outputs, which is an immediate priority of this ongoing research. Despite this, the outputs from our model can provide useful insights into the relevance and potential of spatial microsimulation and dynamic spatial microsimulation in particular for analysing social and spatial inequality in educational attainment. The following section presents some of these outputs, further illustrating this potential.

13.4 Model Results

The dynamic spatial microsimulation model developed was used to analyse social and spatial inequalities in higher education entry and attainment. In this section, we present some model outputs for the city of Sheffield, which is a major city in our study region. In particular, we present the outputs of three events that were modelled: "higher education entry", "graduation" and "employment". Figure 13.4 depicts the geographical distribution of the simulated individuals that "enter" higher education in the first simulation year. Figure 13.5 shows the spatial distribution of

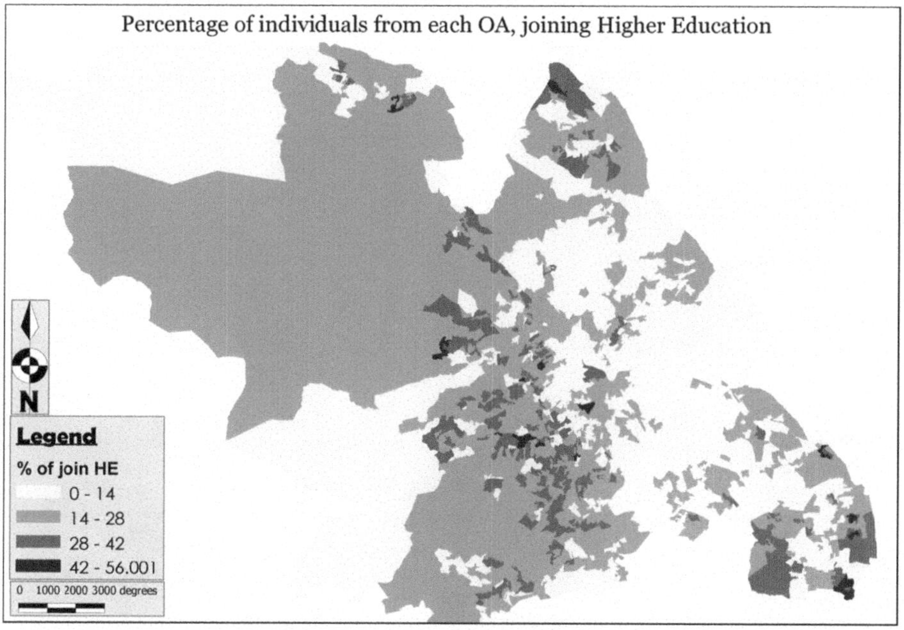

Fig. 13.4 Percentages of individuals joining higher education, dynamic spatial microsimulation

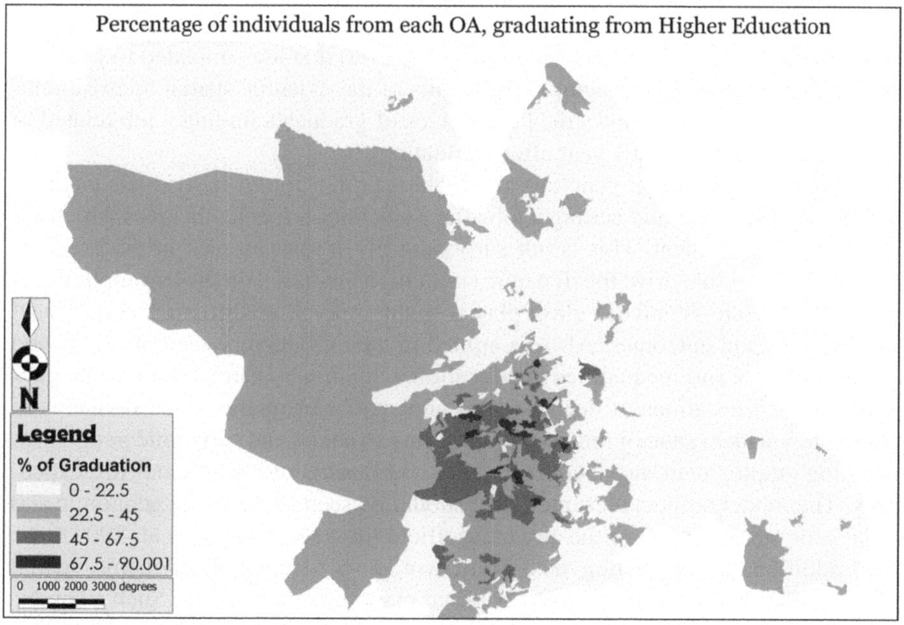

Fig. 13.5 Percentages of individuals graduating from higher education, dynamic spatial microsimulation

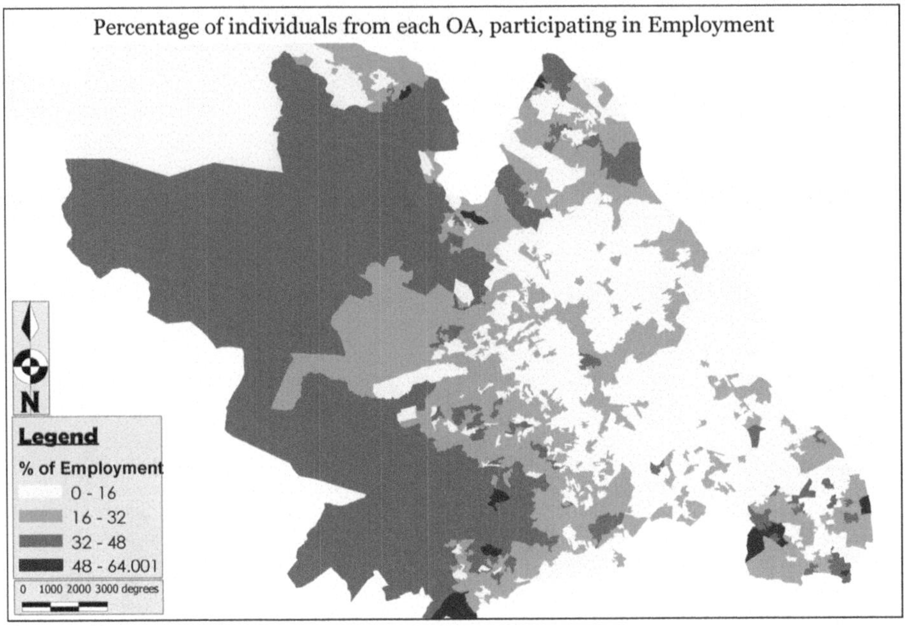

Fig. 13.6 Percentage of graduates finding a job related to the subject studied (1 year after graduation), dynamic spatial microsimulation

the individuals across Sheffield (at output area level) that are simulated to graduate from higher education according to the results of the dynamic spatial microsimulation model. Figure 13.6 shows the percentages of graduates finding a job related to the subject studied within 1 year after graduating.

It should be noted that generally the highest proportion of individuals entering university, graduating and getting a job after university is located in areas which are relatively more affluent. This is not surprising given that, as seen in the previous section, the rules that drive the dynamic simulation are based on the assumption that income and socio-economic class play a major role in determining educational opportunities and outcomes. Also, as argued in the introductory section, social and spatial divisions and inequalities in educational attainment can be seen as the geographical manifestations of deeper social and spatial inequalities and divisions in relation to a wider range of indicators pertaining to social and economic geography, including employment and income, quality of housing and health and life expectancy. The model outputs presented here should be seen in the wider economic and social geography context of the city of Sheffield (also see Thomas et al. 2009).

In addition, an interesting possibility would be to combine dynamic spatial microsimulation modelling with agent-based modelling approaches. Such a combination would involve the replacement of microsimulated units driven by transition probabilities, with adaptive rule-based agents (Williamson 1999).

References

Ainsworth, J. W. (2002). Why does it take a village? The mediation of neighborhood effects on educational achievement. *Social Forces, 81*, 117.

Allen, J., & van der Velden, R. (2001). Educational mismatches versus skill mismatches: Effects on wages, job satisfaction, and on-the-job search. *Oxford Economic Papers, 53*(3), 434–452.

Azzoni, C. R., & Servo, L. M. S. (2002). Education, cost of living and regional wage inequality in Brazil. *Papers in regional science, 81*(2), 157–175.

Ballas, D. (2004). Simulating trends in poverty and income inequality on the basis of 1991 and 2001 census data: a tale of two cities. *Area, 36*(2), 146–163.

Ballas, D., Clarke, Dorling, Eyre, Rossiter & Thomas. (2005a). SimBritain: A spatial microsimulation approach to population dynamics. *Population, Space and Place, 11*(1), 13–34.

Ballas, D., Clarke, G. P., & Wiemers, E. (2005b). Building a dynamic spatial microsimulation model for Ireland. *Population, Space and Place, 11*(3), 157–172.

Ballas, D., Clarke, G., Dorling, D., & Rossiter, D. (2007). Using SimBritain to model the geographical impact of national government policies. *Geographical Analysis, 39*(1), 44–77.

Birkin, M., Clark, G. P., & Clark, M. (1996). Urban and regional modelling at the microscale. In G. Clarke (Ed.), *Microsimulation for urban and regional policy analysis* (pp. 10–27). London: Pion.

Census Dissemination. (2008). *CAS-WEB*. Available at: http://casweb.mimas.ac.uk/. Accessed 15 Aug 2011.

Cheshire, P., & Sheppard, S. (2004). Capitalising the value of free schools: The impact of supply characteristics and uncertainty. *The Economic Journal, 114*, 397–424.

Clarke, C. (2003). *Pupil-centered learning: Using data to improve performance*. London: Department for Education and Science.

Clarke, G., & Langley, R. (1996). A review of the potential of GIS and spatial modelling for planning in the new education market. *Environment and Planning C: Government and policy, 14*(3), 301–323.

Cole, K., & Dale, C. M. A. (1993). The 1991 local base and small area statistics. In *The 1991 census user's guide* (pp. 201–247). London: HMSO.

Demack, S., Drew, D., & Grimsley, M. (2000). Minding the gap: Ethnic, gender and social class differences in attainment at 16, 1988–95. *Race, Ethnicity and Education, 3*(2), 117–143.

Duranton, G., & Monastiriotis, V. (2001). The evolution of the UK North-South divide: Should we mind the gap? *European Investment Bank: Cahiers BEI, 6*(2), 42–57.

Duranton, G., & Monastiriotis, V. (2002). Mind the gaps: The evolution of regional earnings inequalities in the UK, 1982–1997. *Journal of Regional Science, 42*(2), 219–256.

Fryer, R. H. (1997). *Learning for twenty-first century*. London: National Advisory group for Continuing Education and Lifelong Learning.

Gibbons, P. (2002). *Scaffolding language, scaffolding learning: Teaching second language learners in the mainstream classroom*. Portsmouth: Heinemann.

Gibbons, S., & Machin, S. (2003). Valuing English primary schools. *Journal of Urban Economics, 53*(2), 197–219.

Gilbert, N., & Troitzsch, K. G. (1999). *Simulation for the social scientist* (1st ed.). Philadelphia: Open University Press.

Hall, S. (1961). The new student. *Higher Education Quarterly, 15*(2), 152–163.

Harris, R., & Johnston, R. (2008). Primary schools, markets and choice: Studying polarization and the core catchment areas of schools. *Applied Spatial Analysis and Policy, 1*(1), 59–84.

IISER. (2006). *Quality profile: British household panel survey version 2.0*. Available at http://www.iser.essex.ac.uk/bhps/quality-profile/. Accessed 30 Aug 2011.

Kasarda, J. D. (1993). Inner-city concentrated poverty and neighborhood distress: 1970 to 1990. *Housing Policy Debate, 4*(3), 253–302.

Kavroudakis, D. (2009). *Spatial microsimulation for researching social and spatial inequalities of educational attainment*. PhD thesis, University of Sheffield, Sheffield, UK.

López Bazo, E., & Motellón, E. (2009). Human capital and regional wage gaps. *Documents de Treball (IREA), 24*, 1–10.

Miranti, R., Harding, A., McNamara, J., Vu, Q. N., & Tanton, R. (2010). Children with jobless parents: National and small area trends for Australia in the past decade. *Australian Journal of Labour Economics, 13*(1), 27–47.

Pacione, M. (1997). The geography of educational disadvantage in Glasgow. *Applied Geography, 17*(3), 169–192.

Rodríguez-Pose, A., & Tselios, V. (2009). Education and income inequality in the regions of the European Union. *Journal of Regional Science, 49*, 411–437.

Rossiter, D., Ballas, D., Clarke, G., & Dorling, D. (2009). Dynamic spatial microsimulation using the concept of GHOSTs. *International Journal of Microsimulation, 2*(2), 15–26.

Singleton, A. D. (2009). Data mining course choice sets and behaviours for target marketing of higher education. *Journal of Targeting, Measurement and Analysis for Marketing, 17*(3), 157–170.

Taylor, M. F., Brice, J., Buck, N., & Prentice-Lane, E. (2001). *British household panel survey user manual volume A: Introduction*. Colchester: University of Essex.

Thomas, Pritchard, Ballas, Vickers & Dorling (2009). *A tale of two cities: The Sheffield project*. Sheffield: Social & Spatial Inequalities Research Group, Department of Geography, University of Sheffield. Available at http://www.sasi.group.shef.ac.uk/research/sheffield/a_tale_of_2_cities_sheffield_project_final_report.pdf. Accessed 30 Jan 2011.

UCAS. (2008). *UCAS: statistical services*. Available at http://www.ucas.ac.uk/about_us/stat_services. Accessed 15 Aug 2011.

White Rose. (2010). Home. *White Rose University Consortium*. Available at http://www.whiterose.ac.uk/. Accessed 24 Jan 2011.

Wilkinson, R., & Pickett, K. (2009). *The spirit level: Why greater equality makes societies stronger*. New York: Bloomsbury Pub Plc.

Williamson, P. (1999). Microsimulation: An idea whose time has come?. In *39th European Regional Science Association Congress, University College Dublin, Dublin, Ireland* (p. 27). Dublin: University College Dublin.

Chapter 14
Challenges for Spatial Dynamic Microsimulation Modelling

Mark Birkin

14.1 Introduction

It is well known, indeed a commonplace observation, that microsimulation has been established as a technique within the discipline of economics for more than 50 years. It is fair to say that economists have mostly found value in microsimulation as a means to understand the distributional effects of aggregate policies at an individual level. Such questions are essentially static, or at best comparatively static in their approach; they consider perhaps the way a tax regime (static) or the impact of a change in financial policy (comparative static) affects individuals and households within the population. Geographers have also remarked, albeit somewhat less frequently, that the roots of spatial microsimulation also extend right back to the 1950s and the work of Torsten Hagerstrand on migration, innovation and diffusion of technology and ideas. What is notable about this work in the current context is that it introduces from the first not just the importance of spatial disaggregation but emphasises the dynamic processes through which spatial structure evolves.

To the extent that the early approaches of Orcutt and Hagerstrand are distinctive, over time, the work of the economists and geographers has tended to converge. Of course, there are now notable examples of dynamic models in the arena of fiscal microsimulation (such as DynaSim, DynaCan; see, e.g. Morrison 2007) whilst geographers have often been content in the use of static models as a means to disaggregate and refine spatial distributions, for example, as a means to estimate small area variations of income (Birkin and Clarke 1989), health (Smith et al. 2006; Procter et al. 2008; Tomintz et al. 2008) or educational attainment (see Chap. 13 of this book) and as a means to assess the demand for infrastructure services such as water (Jin 2009). The essential point to make is that good examples of dynamic

M. Birkin (✉)
School of Geography, University of Leeds, Leeds, UK
e-mail: m.h.birkin@leeds.ac.uk

R. Tanton and K.L. Edwards (eds.), *Spatial Microsimulation: A Reference Guide for Users*, 223
Understanding Population Trends and Processes 6, DOI 10.1007/978-94-007-4623-7_14,

(spatial) microsimulation are still not especially abundant. One of the objectives of this chapter is to explore why this might be the case, although in essence much of this may be condensed into the simple observation that the methods required are not especially straightforward! We will begin by describing our own approach to the development of a dynamic spatial microsimulation model of the UK population. From this, there follows a discussion of the problems involved in refinement and application of the model. The challenges which arise are of sufficiently general interest, we would argue, as to provide important elements of a research agenda in microsimulation modelling.

14.2 A Dynamic Spatial Microsimulation Model of the UK Population

In this section, we give a high-level description of the structure of a dynamic model ('Moses') which has evolved at Leeds over a period of time and then discuss some of the key features of the approach.

The Moses model has three essential components – a population reconstruction model (PRM), a dynamic projection module and a behavioural simulator.

14.2.1 The Population Reconstruction Model

The PRM recreates a base-year population for a given city or region using a combination of two inputs, both derived from the UK census but at different levels of resolution: household-level data from the sample of anonymised records (HSAR) and neighbourhood data from the small area statistics (SAS). The model creates a complete representation of the national population (UK, Wales and Scotland) on an area-by-area basis. For each household and their component individuals, we represent a wide variety of key socio-economic and demographic attributes, comprising age, gender, marital status, occupation, ethnicity and health status, as well as housing variables including tenure, household size and composition.

Of course, methods for population reconstruction are abundant in the literature. We have experimented with a number of alternative approaches, including synthetic regeneration (Birkin and Clarke 1988; Beckmann et al. 1996), simulated annealing (Williamson et al. 1998; Ballas et al. 2005) and a genetic algorithm (GA) (Williamson et al. 1998; Birkin et al. 2006). If the problem is cast in terms of selecting a subset of individual records from the HSAR to represent the characteristics of each small area, then the GA is superficially attractive as a technique, as we have in essence a binary string (1 for inclusion, 0 for exclusion) which needs to be optimised. However, in practice, the constraints are difficult to represent, the method is computationally extremely expensive, and existing algorithms require significant customisation.

The other two methods, synthetic regeneration and simulated annealing, have both proved to be much more robust. Some experimental results are discussed by Harland et al. (2011) who argue that simulated annealing is a slightly more accurate method, but synthetic regeneration is nevertheless fully satisfactory. We have tended to prefer synthetic regeneration on the basis of its simplicity, the familiarity of the method to the model authors and again the ease by which constraints may be represented, which is especially important given the national scale and scope of this project. Having said this, simulated annealing is a more popular technique but most frequently applied at the level of a single region in which the data processing implications are more limited (see, e.g. Williamson et al. 1996; Smith et al. 2006; Procter et al. 2008). SimBritain is one early example in which simulated annealing has been attempted at a national scale (Ballas et al. 2005).

Individuals in Moses are completely enumerated and attributed to individual output areas within local authority districts, currently amounting to 484 areas in Great Britain. The population can therefore be aggregated at any required scale from the local to the national according to purpose. One of the most useful features of the household SAR is that it comprises both households and their constituent individuals, so that relationships between individuals are preserved. As we shall see below, preservation of these relationships is a significant challenge for the dynamic procedures.

14.2.2 Dynamic Projection Module

The dynamics are represented as key demographic transitions in a series of discrete model steps. Separate modules have been created to represent fertility, ageing and mortality; inter-regional migration; international migration; changing health status; household formation and dissolution (including partnerships and marriage); local migration; and housing market dynamics. The modules are at single-year time intervals up to 25 years in the future (for a more detailed description, see Wu et al. 2008 and Chap. 11 of this book).

The dynamics in Moses are represented as transitions rather than events. The benefits of event-based modelling have been promoted recently by the DynaCan group and others (e.g. Morrison 2007; Gampe et al. 2007), but whilst it is computationally more expensive, the transitions approach is much simpler to both implement and understand and is well suited to the data structures for this project, which are mostly based on aggregated transitions data rather than individual demographic events. Thus, for example, fertility is measured by the number of births to mothers in a single-year period according to their age and marital status, so it makes a lot of sense to represent the process as a transition (to parenthood) over a single year in the model. The parameterisation of processes is richly specified; for example, in the ageing module, survival probabilities are modelled for single years of age by gender for each of 33 administrative wards (for which vital statistics are available) resulting in a total of 6,666 survival rates for this single local authority area.

Whilst the list of demographic modules is quite extensive, it can be seen that not all of the attributes enumerated within the PRM are explicitly accounted for by the dynamic model at this stage. In particular, characteristics such as education, occupation and car ownership are not directly impacted by any of the transition modules. Therefore, one of the most important challenges is to extend the functionality of the Moses dynamic model so that transitions between education and age categories are modelled explicitly. One of the most useful resources to do this is the British Household Panel Survey (BHPS) in which a cohort of approximately 15,000 UK households has been tracked through time since 1991 (Taylor et al. 2005). The BHPS already provides a major input to the process of household formation and dissolution (Birkin et al. 2009a, b) and can be used to identify movers according to characteristics such as household size, marital status, age and ethnicity. In the same way, the problem of modelling transitions between educational states or from one occupation to another is currently underway as part of a broader study into the relationship between educational attainment, employment and life chances (Warner et al. 2010; Lambert and Birkin 2012).

14.2.3 Behavioural Models

Whilst demographic change is both interesting and fundamental, many of the applications to which Moses is directed require consideration of a much wider array of attributes relating to transport, health, crime and so on. In order to extend the range of the system, a synthetic 'linker' has been developed which allows individuals in the simulation to be linked with (demographically) similar people in survey datasets such as the National Travel Survey or Hospital Episode Statistics. Using this linker, we can represent the fact that more elderly individuals are much more likely to require treatment for, say, cataracts or that large households are more likely to exhibit high levels of car ownership.

An example at this point may be helpful. Suppose that we are interested in changing patterns of smoking over a 25-year period. Starting with the PRM, we can link to records in a dataset which captures individual variations in the propensity to smoke according to key characteristics such as age and 'social class' (occupation). Then we take the dynamic version of the simulation 25 years into the future and do exactly the same thing. For example, if the population in a given area becomes much more elderly, and assuming that young people are the heaviest smokers, then other things being equal the ratio of smokers in our area of interest will be reduced.

An obvious objection to this procedure is that in effect, we assume no change in the behaviours represented within our underlying dataset over a 25-year period, but this can be readily assessed. For example, one could quite easily adjust the model to assume instead that there will be a 1% annual reduction in smoking for the next 25 years, or that smoking rates between the genders will equalise, or whatever other assumptions or scenarios are seen as appropriate (and, whilst these scenarios may be far from straightforward, it is more a question for the application context and the problem towards which the simulation is directed than for the method itself).

A more fundamental problem is a conflict between this (naive) approach to behavioural linkage and the dynamics of the simulation. Suppose, for example, that at the start of the simulation individual x, a male aged 25, is a smoker and individual y, a female aged 40, is a non-smoker. After 25 years, we run an independent linkage operation, and individual x, now aged 50, is a non-smoker and individual y now turns out to be a smoker. Whilst the net effect is the same, it seems much more likely that individual x would still smoke and individual y still would not. This may not be important if the ultimate objective of the analysis is just to estimate the number of smokers by demographic type or small geographical area (as could be the case with many of the static and comparative static approaches which we mentioned in our introduction) but could be more significant if we want to look at dynamics explicitly, for example, if we introduce a new policy or stop smoking service at time t and want to see the effect of this. The problem here is similar to the approach which we have sketched in relation to education and income in the previous subsection. So in essence, what we would like to do is to represent the transition between categories (e.g. from smoker to non-smoker) as a function of relevant demographic or socio-economic variables, using appropriate data from BHPS or health surveys. A generalisation and extension of this method might take further account of the importance of interdependence and environmental interactions in the decision-making process. For example, the decision to stop smoking could be related to the fact that a friend or relation has already changed their behaviour. Alternatively, the decision to become a car owner could be dependent on a recent change of house or job, or moving to a location with more difficult public transport links. For problems of this type, the technique of agent-based modelling has significant potential, and this is pursued further in the examples and discussion which follow.

14.2.4 Towards a Social Simulation Infrastructure

In the work which has been undertaken on Moses as part of the UK e-Social Science programme (Birkin et al. 2009a, b), a multidisciplinary team from a variety of institutions has also considered the problem of how to embed the simulations within a research infrastructure which makes these simulations easy to use, share and interpret (Birkin et al. 2010). A proposed 'architecture' for this work is reproduced in Fig. 14.1. Here, we can see that the basic modules already discussed are embedded within a service-oriented framework which makes these capabilities – reconstruction, dynamics and linkage – easy to access and combine. In addition, the important service of visualisation has been added, to allow for easy extraction and representation of key pieces of information, through maps or other graphical means. Most important of all, however, Fig. 14.1 emphasises the need for the *engagement* in this process of planners, academics, policymakers and research users.

We will now move on to discuss various ways in which we see dynamic microsimulation tools providing substantial value to applied problems.

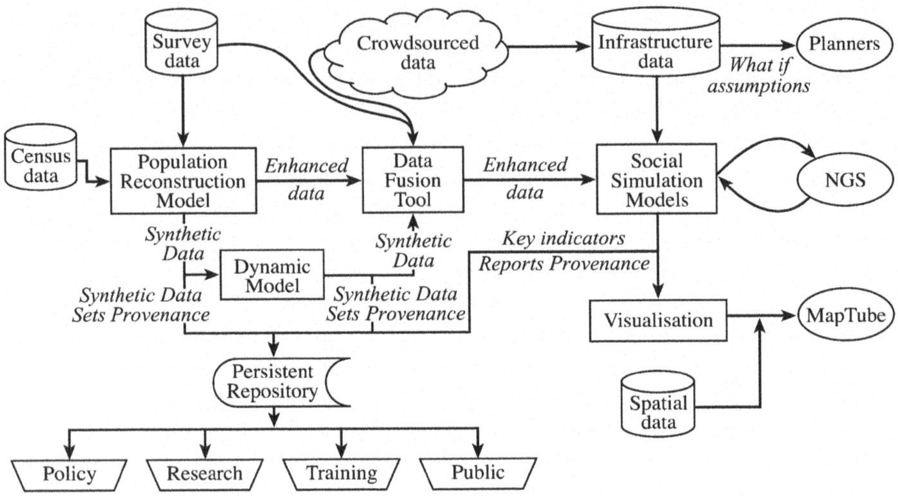

Fig. 14.1 An architecture for dynamic simulation

14.3 Applications

In this section, several applications of the dynamic spatial microsimulation technology are reviewed. Through these examples, it is possible to add substance to the argument that important problems can be addressed in this way. These case studies also serve to highlight some of the more important difficulties and issues as a precursor to the discussion of research challenges which follows in Sect. 14.4.

14.3.1 Demographic Modelling, Health Care and Social Service Provision

A straightforward exploitation of the dynamic model is to explore future scenarios for health care and social services based on demographic change. In collaboration with the Deputy Director for Social Services, Leeds City Council, a needs assessment for social care was produced, and although more sophisticated modelling has some value, much of this was achievable using the power of the basic approach. The essential problem here is that service providers have good information about the uptake of services, but find it much more difficult to understand how this may be related to unmet requirements. The problem of how to estimate future changes in both need and uptake is also significant. So one important capability here is simply to forecast the changing pattern of key indicators into the future. The example of limiting long-term illness is shown in Fig. 14.2 (whilst comparable Australian projection attempts include Harding et al. 2011).

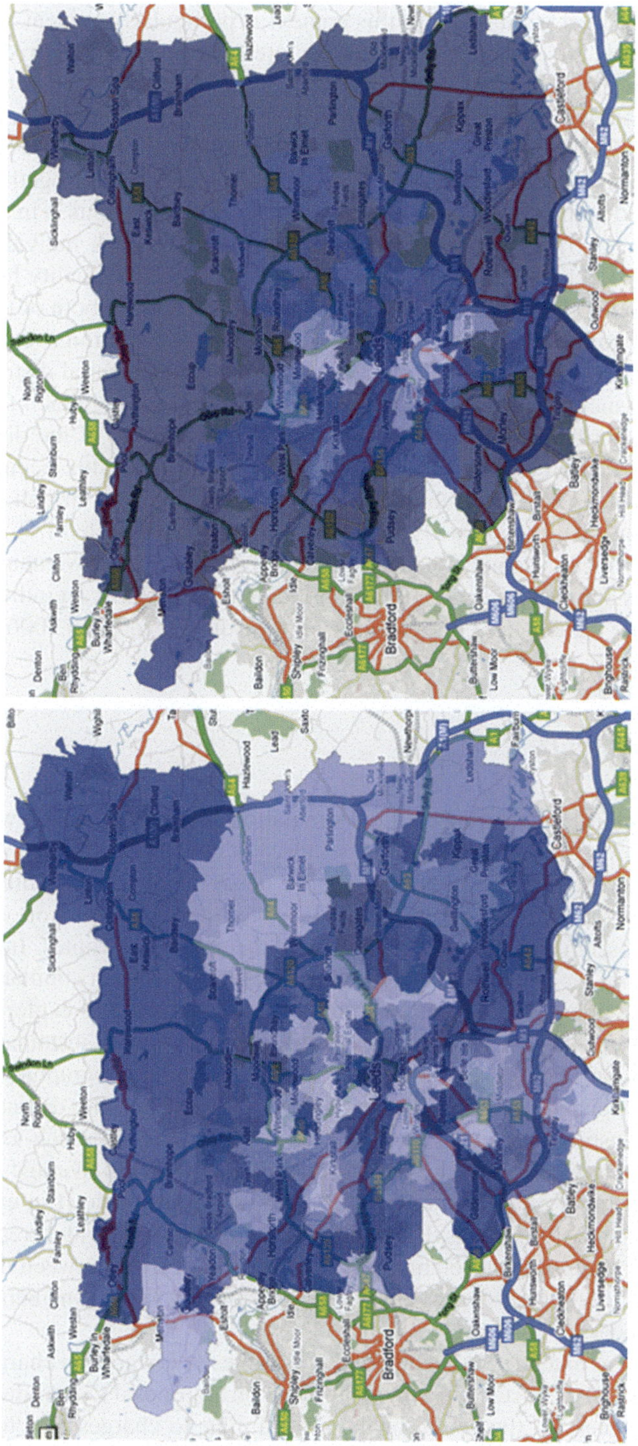

Fig. 14.2 Limiting long-term illness, 2001 and 2031

The value of microsimulation in this context is particularly evident since social care requirements are often dependent on the intersection of multiple risk factors at the level of an individual person or household. These numbers cannot be simply extracted from aggregate census information. For example, one high-risk group which can be identified in this way are codependent adults. These were defined in the simulation as two elderly people (aged 65 and over) living together with no other household members and in which at least one of the residents is in a poor state of health (Fig. 14.3). In a similar way, we identified high-risk individuals living alone as not just elderly (65 or over), in poor health, with low-quality housing and with poor mobility due to lack of access to a car (Fig. 14.4). In both of these cases, it was possible to estimate current distributions and their evolution over time.

To complement this approach, we were able to estimate spatial distributions of key indicators such as disability through an extension to the approach for smokers which were described in the previous section. Thus, a profile of the disabled was obtained – in relation to age, household status, quality of health, occupation and employment. Measures of disability were generated from the British Household Panel Survey and then extrapolated 25 years into the future (see Table 14.1). A scenario-based projection was also generated assuming a single-year improvement in healthy life expectancy for every 5 years in the simulation (e.g. in 2031, an individual aged 65 has the same incidence of disability as an individual aged 60 in 2006).

14.3.2 Transport

In order to demonstrate the power of the simulation architecture which we sketched in Fig. 14.1, a case from the transport sector was developed. The vital extension which facilitates this application is a simple behavioural model which can take detailed demographic inputs from either the baseline reconstruction or dynamic model and to assign transport destinations and the associated routes. In joint work with the Institute for Transport Studies at Leeds, the spatial microsimulation models were loosely coupled with Omnitrans route planning software to explore the effect of population change on accessibility, traffic congestion and sustainability. This operation is simple and essentially sequential. First, the demographic projections are produced. Second, these projections are used to generate the demand for trips within the transport model. A sample output from this approach is shown in Table 14.2 where we can see the growth in trips, reduction in average speeds and increase in airborne pollutants over a 25-year projection cycle (Liu et al. 2009). The significance of this work is that traffic simulation tools are widespread and have sophisticated capabilities in relation to the allocation of trips between destinations and the distribution of trips between routes, but typically lack refinement in the trip generation process.

This work was extended in a study of the Manchester congestion charge. In 2008, the residents of Manchester were invited to vote on a proposal to introduce a charge on vehicular access to the centre of the city, similar to charges which have been

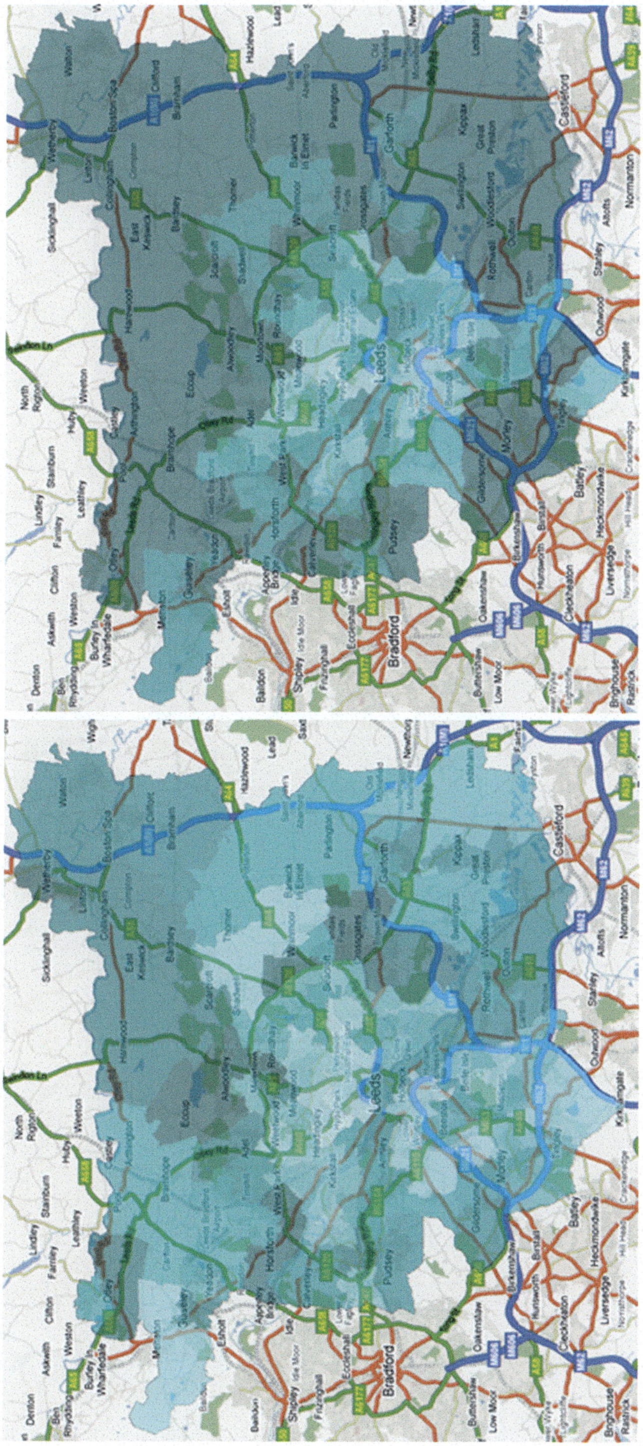

Fig. 14.3 Codependency, 2001 and 2031

Fig. 14.4 Deprivation, 2001 and 2031

Table 14.1 Projections of disability for the city of Leeds

Year	Estimate	Disabled (%)	Comment
2006	Baseline	9.1	Disability rates from the BHPS, 2006
2031	Baseline	14.1	Continued disability rates from 2006
2031	Scenario	11.0	Enhanced healthy life expectancy

Table 14.2 Changing travel patterns in the city of Leeds

	2001	2006	2011	2016	2021	2026	2031
Population	680,872	696,778	702,290	720,802	771,918	801,447	820,114
Average speed (km/h)	52.9	52.1	51.1	49.7	48.1	46.9	46.0

operational in central London since the earlier part of the decade. In order to gauge public opinion on this issue, our project partners at the Centre for Applied Spatial Analysis (CASA) at University College London were invited to prepare a web-based survey of attitudes and responses to the charge. This survey was promoted via regional television (BBC North-West) and received over 15,000 responses. To exploit this data, we constructed a simple traffic simulation tool which was embedded within the dynamic simulation architecture. The trip patterns produced by this model were conditioned by parameters relating to travel times, mode preferences, trip cost and so on (for more details, see Birkin et al. 2011a). For a given set of parameter values, then for any combination of origin characteristics, it is possible to extract from the traffic model probabilities of selecting any destination or route combination. These combinations can be sampled in the usual way in order to attach a destination and route for each individual in the microsimulation (i.e. by Monte Carlo selection).

The procedure was as follows: A dynamic microsimulation was run in order to update the population to the present day (i.e. 2008 at the time of the study). The traffic simulation was run with a given parameter set. From the traffic simulation, destination and route choices were appended to the microsimulation. This could then be used to predict the transport behaviour of the individuals in the MSM. Then the same process was repeated assuming the introduction of a congestion charge, which naturally affects the choice of both destinations and routes quite significantly. From this 'scenario', we can gauge the effect of a congestion charge on individual behaviour. What we did next was to run a huge number of simulations in order to reproduce the stated behaviours from the congestion charge survey within the MSM. This required the use of a genetic algorithm to guide the search procedure, and the computational intensity of the process was supported by deploying the models across a national grid infrastructure (Birkin et al. 2011a). Having achieved all this, it was then a straightforward exercise to use the calibrated parameters to test the effect of alternative charging scenarios on the transport behaviours of the local population

Table 14.3 Travel behaviour under alternative policy scenarios in the Manchester congestion charge model

	Survey	Scenario 1	Scenario 2	Scenario 3	Scenario 4
		33 wards in central ring with £5 charge	33 wards in central ring with £10 charge	69 wards in extended ring with £5 charge	69 wards in extended ring with £5 charge
Carry on (%)	7	9	2	10	2
Change route (%)	9	6	4	5	2
Change mode (%)	7	7	12	12	22
Change destination (%)	23	23	27	24	30
Unaffected (%)	54	55	55	49	44

Survey data is calibrated against a dynamic simulation model to reproduce five behavioural responses following the introduction of a congestion charge. Four alternative scenarios are then considered with alternative spatial configurations and charging levels (some of the key outputs from the simulation tool are shown in Table 14.3) although a much wider range of indicators were also produced including effects on distance travelled, trip costs, pollution and road accidents (see also Birkin et al. 2010).

In this example, we have therefore introduced primary survey information (increasingly referred to as 'crowd-sourced data') in order to calibrate the behaviour in a microsimulation model, and in which the behavioural estimation process is somewhat more sophisticated than the simple linkages with which we began in Sect. 14.3.1. This process could potentially allow policymakers to explore the impact of alternative planning scenarios but might also be used more widely as an information tool to allow members of the public to gauge the effect of their own choices on the urban environment. However, a disappointing but necessary post-script to this particular case study is that the proposal to introduce a congestion charge in Manchester was comprehensively rejected by the public vote before the end of 2008.

14.3.3 Crime

Spatial analysis of crime patterns is of great interest not just to academics (Eck 1995) but to a wide variety of groups in both the public sector (e.g. police) and private organisations (insurance companies and so on) (Hirschfield and Bowers 2001). Studies have persistently identified local hotspots in the concentration of incidents which are relevant to both understanding the underlying behaviour patterns by which criminal activity is realised as well as having implications for resource targeting, for example, through neighbourhood policing or target hardening. In addition to map-based analyses of criminal incident data, geodemographic analysis has been deployed to try and gain an improved analysis of spatial variations. More promising still are recent agent-based models of criminal activity; for example, Malleson (2010) has constructed a model of burglaries which is predicated on a detailed individual-level simulation of the attributes and activity patterns of burglars. This model also includes a detailed assessment of opportunities relating to the characteristics of individual neighbourhoods and streets (such as quality of lighting, ease of access, vacant dwellings, etc.). None of these approaches, however, allow detailed assessment of the importance of victims within the criminal environment.

Environmental criminology shows that the attributes and behaviour of victims are of crucial importance, in relation to the occupancy of properties (e.g. at different times of day), the attractiveness of targets and the propensity to repeat victimisation, which are all strongly related to the characteristics of individual households. Recent work (Malleson and Birkin 2011) has demonstrated how to couple an agent-based model of burglary with a microsimulation model of the targets (i.e. potential victims). An ongoing project funded by the Joint Information Services Committee (JISC) in partnership with such agencies in Merseyside, Manchester and Leeds is seeking to demonstrate and exploit the value in this approach (JISC 2011). Whilst it is true that synthetic individual data has little direct benefit, the flexible aggregation of these data, for example, to the level of streets or small estates, would facilitate the forensic examination of patterns such as hotspots according to the social make-up of local communities alongside observable physical and environmental criteria. The systematic linkage between synthetic microdata and data about offences (from police or local authorities) or about victimisation rates from survey data such as the British Crime Survey, together with diverse information about local environments, could therefore furnish much valuable intelligence to both criminologists and local agencies.

14.3.4 Housing

Local housing markets have a fundamentally important place in the Moses dynamic model by conditioning intra-regional migration flows. An understanding of housing and residential location patterns is therefore of vital importance to the modelling process; it is also capable of yielding scenarios which are important to policymaking in their own right. In a current doctoral research project which has been supported by the Chief Regeneration Officer in the city of Leeds, we have considered how to represent housing market decisions in order to assess the effects of changes in the supply of accommodation (e.g. a continually changing balance between public and private providers) as well as broader influences such as the quality of local schooling and the nature and availability of local employment. All of this has been grounded within the context of EASEL (East and South East Leeds), a project which aims to regenerate one of the more deprived communities in Leeds through a combination of physical, social and economic redevelopment (Jordan et al. 2011).

Through a combination of literature review and examination of empirical evidence, local migration patterns have been assessed in relation to a simplified two-stage process: first, a decision to move and second, a choice of destination. Major factors in the decision to move are household composition and life stage, tenure, socio-economic status and accommodation type (Jordan et al. 2011, pp. 157–158). In relation to the destination choice process, seven major rules were identified relating to proximity, housing type, tenure, ethnicity, transport, local schooling and neighbourhood quality (ibid., p. 160). Specific scenarios are considered by Jordan (2011) in relation to changes in the housing stock (in particular, investment through ‘public-private’ partnerships for new builds and conversion of existing properties), a proposal for a new road scheme to provide better connections to the major employment centres of central Leeds and alternatives for the closure or revitalisation of a local school.

A second application of spatial MSM concerns current changes in government policy relating to housing benefit. Recent proposals are both extensive and complex but affect both eligibility and levels of benefit as part of the drive to cut £1.8 billion from the national budget (Birkin et al. 2011b). The local impact of these changes is important, partly in view of potential interactions with other dimensions of social and economic deprivation, but also because housing markets (whether rental or owner-occupied) are themselves localised and it seems unlikely that such large sums of money can simply be removed from the system without knock-on effects to the quality and availability of property (i.e. some understanding of the reaction of private landlords and housing providers is also required). Through ongoing research with Leeds City Council, we hope to shed light on the immediate geodemographic implications and effects of the housing benefit reforms; to understand the likely longer-term market dynamics and outcomes in relation to both local neighbourhoods and the city; to identify how these changes and their effects may interact with other austerity measures, policy changes (e.g. to social rented housing) and economic conditions across Leeds; and to explore issues of social and spatial justice and equality in relation to these reforms, for example, how they might have specific implications for different social groups and communities across the city.

14.3.5 Infrastructure

The potential value of simulation in planning has already been considered in the context of transport and housing. Other services which can potentially benefit from the same treatment include energy, water, waste disposal and information and communication technologies. Whilst there is some tradition in the assessment of water consumption through a microsimulation approach (e.g. Williamson et al. 1996; Jin 2009), the other services have received less attention in the past. A current project with the Infrastructure Transitions Research Consortium (ITRC) aims to utilise microsimulation as a means to embed demographic change as a significant factor underpinning long-term demand for infrastructure services (water, waste, ICT, transport and energy – ITRC 2011). The ITRC is a research grouping with academics from six universities but also includes supporters from many business sectors, including engineering, utilities and insurance, as well as supporters from government and regulatory groups. In this context, long term could mean anything up to the middle of the next century whilst recognising the uncertainties over such a timescale in provision and consumption of these services.

As a preliminary demonstration of capability, dynamic models for a coarse 11-zone model (ten standard regions of England and Wales, plus Scotland) were aligned to the ONS sub-national forecasts to 2033 (ONS 2010) (see Birkin and Wu (2009) for a discussion of methods). In parallel to this, we conducted analysis of consumption patterns for ICT and transport using data provided from Acxiom's Research Opinion Poll (see Thompson et al. 2010a) Four demographic drivers were arbitrarily selected – age, household size, ethnicity and income. For transport, we built an index based on the question 'how many miles did you drive in your own car last year', and for ICT, we repeated this using a combination of ownership of mobile phones, digital television and access to broadband. The results are summarised in Figs. 14.5 and 14.6, showing patterns that might generally have been expected ex ante. In both cases, high large income and household sizes are strongly associated

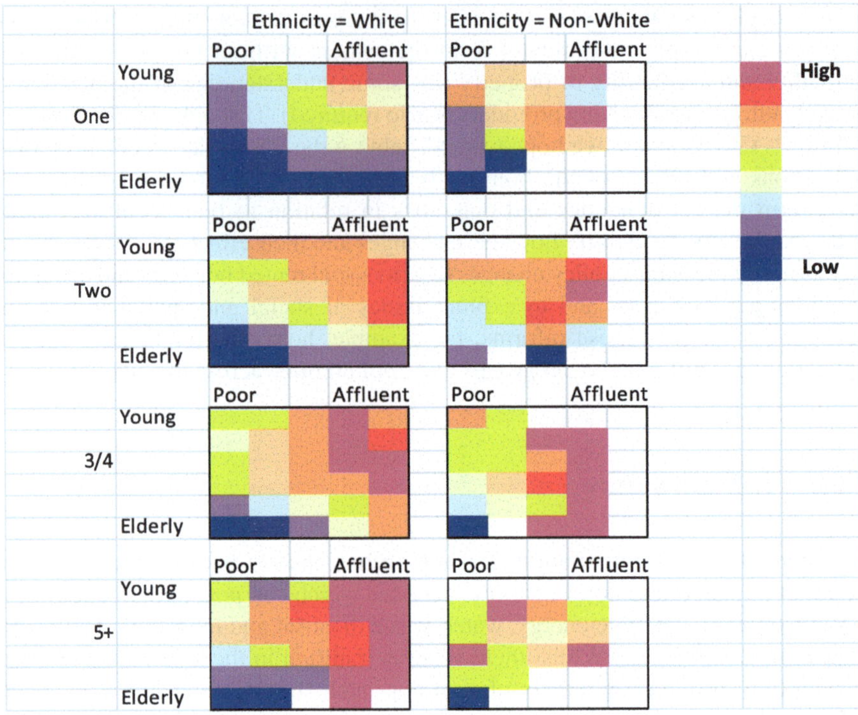

Fig. 14.5 Demand for ICT

with intense consumption. The peak ages for transport use appear to be in the 'family' life stages (i.e. early middle age, 30–45), whilst for ICT, this is slightly younger. Ethnic variations are not tremendously significant.

The forward projection of these data is shown in Figs. 14.7 and 14.8. The baseline assumption is no change in the socio-demographics of consumption. Here, the major structural factors are a combination of differential regional growth, mostly biased towards the south and east, and ageing of the population. In general, demographic growth is offset to some degree by ageing of the population, since consumption rates tend to decline somewhat amongst the elderly. For example, in Great Britain as a whole, an increase of 22% in the number of people yields a 14% growth in the index of consumption for transport and 12% for ICT.

These early explorations obviously contain an alarming array of restrictive assumptions. More refined analysis will concentrate on the adoption of alternative scenarios for both demographic change and consumption patterns and spatial disaggregation of the estimates to a more appropriate level of geography. However, given that the most acute stresses on infrastructure provision are on services like water, waste and transport (for which 'national grids' are not feasible), then the coincidence of these regions as the zones of greatest expansion in demand is immediately thought-provoking for service providers.

Fig. 14.6 Demand for motoring

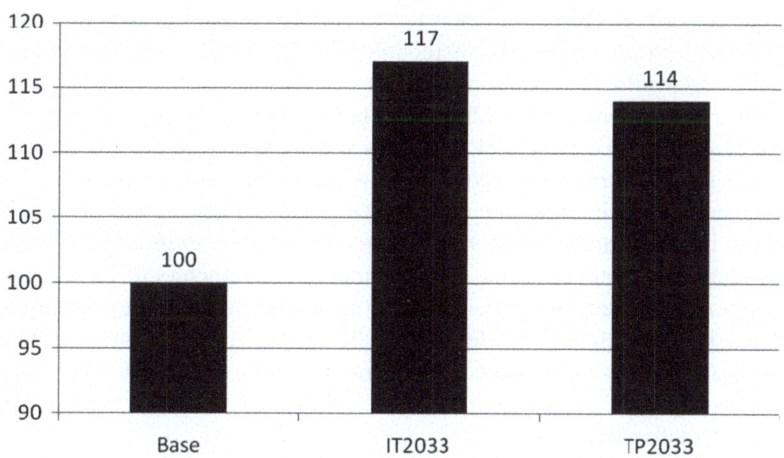

Fig. 14.7 Growth in consumption, 2001–2033

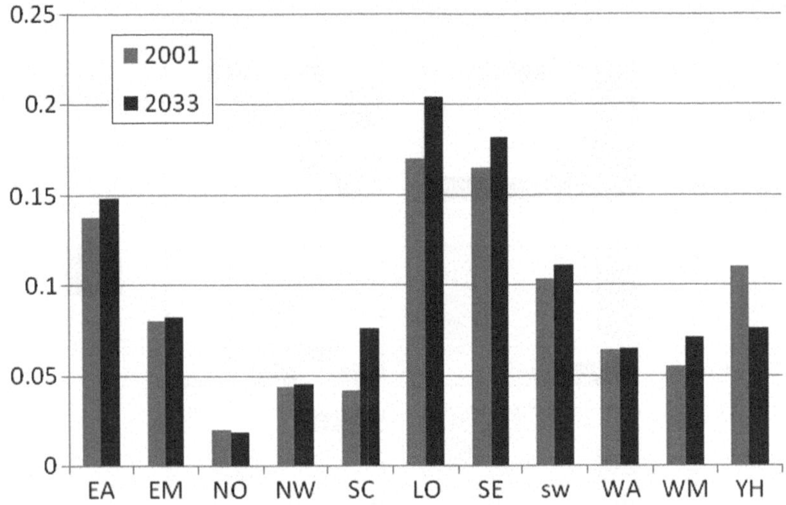

Fig. 14.8 Growth in consumption by region

14.4 Challenges for Dynamic Spatial Microsimulation

The examples which have been described in the previous section are unapologetically ambitious. They stretch the limits of what can be achieved with current technologies right up to and in most cases significantly beyond its current limits. Were they not to do so, then these problems would have little to attract the attentions of academic research. In this section, some of the most important dimensions of the challenge are assessed.

In any modelling exercise, the difficulties of *validation* are rarely to be discounted. This issue is particularly acute when the underlying model is dynamic, looking forward to an uncertain future. As a methodological device, some form of *backcasting* may be the best available validation tool. The general idea, expressed at its simplest, is to run the simulations backwards in time rather than forwards, with the obvious and significant benefit that model outcomes can then be evaluated against something that is known rather than unknown. This technique has been pioneered within climate change science with somewhat mixed results. Sceptics will always claim, not without justification, that historic analyses will do little to inform the trajectory of key system parameters into the future. Since the parameters of local economic and demographic systems may be slightly more stable and subject to regulation and control to a degree, then these methods could be of more interest in microsimulation. The case has at least been broached by Birkin and Malleson (2010) who suggested that backcasting at the very least provides a robust test of model consistency and also offer a tentative example.

Naturally, the future will always be uncertain. The most popular response in microsimulation modelling has been to adopt a procedure of *alignment* to some higher level or aggregate estimates; the alignment of individual-based estimates to government demographic projections, which we introduced briefly in Sect. 14.3.5, is a representative example here. Such approaches tend to emphasise the usefulness of MSM in producing fine-scale disaggregations of macroeconomic forecasts. They do less to exploit the fact that a dynamic MSM might be a better way to go about the forecasting problem in the first place. In other words, can the macroeconomic forecasts themselves be trusted? The answer to this question presumably turns to some degree on the quality of the estimates in question. An alternative strategy based on the formulation of scenarios has been adopted in the context of medium to long-term demographic projections in the UK context (Wohland et al. 2010) and as part of a pan-European study (de Beer et al. 2010). Such approaches can provide a lot of flexibility in 'what if?' planning, without necessarily being highly prescriptive about what might actually be expected to happen.

Academics could almost certainly learn much about the reliability of the techniques, as well as their value, through a process of greater *engagement* with planners, policymakers and end users, although to be fair, microsimulation has tended to focus much more closely on real-world problems than the competing method of agent-based modelling. This argument was put forward in the plenary address at the second IMA conference (Gilbert 2009) and in the literature by Wu et al. (2008). Nevertheless, continued and greater engagement provides a number of benefits.

First is the possibility of real data with which to calibrate, test and enhance the models. In the examples above, we have been provided with data on housing schemes and tenant attitudes and behaviour relating to the EASEL scheme, a complete anonymised list of housing benefits claimants over a 3-year period, intelligence relating to the provision and uptake of social services for Leeds primary care areas as well as commercial data from Acxiom for research applications. We have a long-standing relationship with Safer Leeds who have shared data relating to property theft and other non-violent crimes, by time, season and location (Thompson et al. 2010b). In the fullness of time, we are confident of extracting good information from the utilities about patterns of consumption in order to further inform the infrastructure models.

Second, engagement with real-world users plays to the crucially important impact agenda, which has been a steadily important theme amongst research councils and envisaged to become equally significant in our teaching as the cost of tuition continues to rise and relevance rises ever higher up the agenda.

Finally, the possibility of direct financial support from third parties should not be discounted, as the provision of intelligence to third-party users, in both commercial and public sector organisations, has considerable value (Birkin et al. 2002).

In addressing the difficulties posed by really hard practical problems, researchers have not helped themselves by a tendency towards reinvention. In the related field of geodemographics, Dan Vickers has argued that research has been held back because of the need for commercial users to protect their methods. His solution to this is to promote 'open geodemographics', in which data and techniques are made transparently available to all interested parties (Vickers 2007). Whilst commercial concerns are not usually to the fore in MSM, there seems to be a great deal of duplication and little agreement on the best techniques to use, for example, in the creation of micro-populations, let alone any move towards something like a standard set of simulations. Perhaps the time has come to start thinking seriously about 'open microsimulation'. Certainly, we would see our NeISS project as a nudge in this direction, with its emphasis on the sharing of models and data sources and the publication of algorithms, model results and the 'workflows' which were used in their creation.

A second characteristic feature of academic research is the maintenance of intellectual 'silos'. An important manifestation of this in the current context is a general failure to bring closer together the closely related fields of MSM and *agent-based modelling* (ABM). The two have much in common, if only at the level of system representation. Of course, it can be argued with much justification that the philosophy of the two approaches has some importance differences – on the one hand, typically stochastic (MSM), the other more obviously process-based and behavioural (ABM); one quite highly policy-focused (MSM), the other more abstract (ABM); and so on. We would argue that this provides all the more reason to seek synergies through common ground. In the work reported here, we have looked for the possibilities of integration, especially in demographic modelling, where agent mechanisms have provided for model extensions to deal with difficult problems in local housing markets, student migration and the significance of individual and family histories (Wu et al. 2008, 2011; Jordan et al. 2011). Our work on crime has also begun to explore this territory, as reported above (Malleson and Birkin 2011), and work on agent-based retail models is also beginning to point in this direction (Birkin and Heppenstall 2011). One way forward could be to build on the open microsimulation concept to push the integration of approaches, and in current work, we are evaluating the possibilities of embedding RePast, an ABM software development framework, within the NeISS simulation architecture of Fig. 14.1.

Another disconnection between MSM and ABM is that whilst the former are almost always strategic in their intent, ABM are much more likely to have a real-time emphasis. This is true in applications such as pedestrian flow and crowd control (e.g. Helbing et al. 2005) whilst Kai Nagel's model of vehicular flow in Switzerland has been operational in real time for several years (Nagel et al. 1998). Whilst the inputs to such agent-based models are typically via sensor networks (which in themselves are rapidly becoming more widespread and therefore of increasing value as sources), the work on SurveyMapper and our own transport example raises intriguing possibilities about the possibilities of *crowd-sourced MSM*, which could be operational in either real time or much longer policy and planning timeframes. At the strategic extreme, however, the importance of higher level interdependencies should not be discounted. An important dimension of the ITRC project not yet discussed will be to examine the *co-evolution* of infrastructure systems in relation to both supply and demand. This is no less an issue in the other application sectors; for example, whilst local migration is responsive to the supply of housing, in the medium term, property markets should go some way to ensuring that new housing development begins to take place where it is most needed.

14.5 Conclusion

In this chapter, we have presented a technical and architectural framework for dynamic spatial microsimulation. A range of examples have been used to illustrate the power and scope of interests that can be addressed within this domain. The major challenges in our own research agenda have also been brought forward. We hope to have done enough to convince the reader that this agenda is by nature not an ideal wish list which will never be realised but more a set of tangible objectives against which concrete progress is already being made. As long as academic research can be mobilised to meet these challenges, and others as they arise, then dynamic spatial microsimulation can provide important models to aid our understanding of socio-economic and demographic problems, as well as continuing to provide useful tools for their amelioration through policy and planning.

References

Ballas, D., Clarke, G., Dorling, D., Eyre, H., Thomas, B., & Rossiter, D. (2005). SimBritain: A spatial microsimulation approach to population dynamics. *Population, Space and Place, 11*(1), 13–34.

Beckmann, R., Baggerly, K., & McKay, M. (1996). Creating synthetic baseline populations. *Transportation Research, 30A*(6), 415–435.

Birkin, M., & Clarke, M. (1988). Synthesis: A synthetic spatial information system for urban modelling and spatial planning. *Environment and Planning A, 20*, 1645–1671.

Birkin, M., & Clarke, M. (1989). The generation of individual and household incomes at the small area level using synthesis. *Regional Studies, 23*, 535–548.

Birkin, M., & Heppenstall, A. (2011). Extending spatial interaction models with agents for understanding relationships in a dynamic retail market. *Urban Studies Research*. Article ID 403969, 12 p. doi:10.1155/2011/403969

Birkin, M., & Malleson, N. (2010). *An investigation of the robustness of a dynamic microsimulation model of urban neighbourhood dynamics*. Paper presented to the North American Regional Science Council (NARSC), 10–13 November, 2010, Denver, CO. Available online at http://www.geog.leeds.ac.uk/fileadmin/downloads/school/people/academic/m.birkin/backsim_paper_v2.pdf. Accessed 16 Aug 2011.

Birkin, M., & Wu, B. (2009). *A hybrid spatial microsimulation model for decision support in demographic planning*. IMA: Ottawa.

Birkin, M., Clarke, G., & Clarke, M. (2002). *Retail geography and intelligent network planning*. Chichester: Wiley.

Birkin, M., Turner, A., & Wu, B. (2006, June). *A synthetic demographic model of the UK population: Methods, progress and problems*. Proceedings of the Second International Conference on Esocial Science, Manchester, UK.

Birkin, M., Townend, P., Turner, A., Wu, B., & Xu, J. (2009a). MoSeS: A Grid-enabled spatial decision support system. *Social Science Computing Review, 27*(4), 493–508.

Birkin, M., Wu, B., & Rees, P. (2009b). Moses: Dynamic spatial microsimulation with demographic interactions. In A. Zaidi, A. Harding, & P. Williamson (Eds.), *New frontiers in microsimulation modelling* (pp. 53–78). Aldershot: Ashgate.

Birkin, M., Procter, R., Allan, R., Bechhofer, S., Buchan, I., Goble, C., Hudson-Smith, A., Lambert, P., & de Roure, D. (2010). The elements of a computational infrastructure for social simulation. *Philosophical Transactions of the Royal Society Series A, 368*, 3797–3812.

Birkin, M., Malleson, N., Hudson-Smith, A., Gray, S., & Milton, R. (2011a). Calibration of a spatial interaction model with volunteered geographical information. *International Journal of Geographical Information Science, 25*(8), 1221–1239.
Birkin, M., Clarke, G., Hodkinson, S., & Thompson, C. (2011b). *Credit crunch Britain: Analysis with spatial microsimulation*. Stockholm: IMA.
De Beer, J., Van der Gaag, N., Van der Erf, R., Bauer, R., Fassmann, H., Kupiszewska, D., Kupiszewski, M., Rees, P., Boden, P., Dennett, A., JasiDska, M., Stillwell, J., Wohland, P., De Jong, A., Ter Veer, M., Roto, J., Van Well, L., Heins, F., Bonifazi, C., & Gesano, G. (2010). *DEMIFER: Demographic and migratory flows affecting European regions and cities, Applied research 2013/1/3, final report | version 30/09/2010*. Luxembourg: ESPON.
Eck, J. (1995). Crime places in crime theory. In J. E. Eck & D. Weisburd (Eds.), *Crime prevention studies, volume 4*. New York: Criminal Justice Press.
Gampe, J., Zinn, S., Willekens, F., & van den Gaag, N. (2007). Population forecasting via microsimulation: The software design of the MicMac project. In *Eurostat: Methodologies and working papers; Theme: Population and social conditions* (pp. 229–233). Luxembourg: Office for Official Publications of the European Communities.
Gilbert, N. (2009). *Microsimulation and agent-based modelling* (Plenary address). Ottawa: International Microsimulation Association.
Harding, A., Vidyattama, Y., & Tanton, R. (2011). Demographic change and the needs-based planning of government services: Projecting small area populations using spatial microsimulation. *The Journal of Population Research, 28*(2–3), 203–224.
Harland, K., Birkin, M., Heppenstall, A., & Smith, D. (2011). Creating realistic synthetic populations at varying spatial scales: A comparative critique of microsimulation techniques. *Journal of Artificial Societies and Social Simulation, 15*(1), 1.
Helbing, D., Buzna, L., Johansson, A., & Werner, T. (2005). Self-organized pedestrian crowd dynamics: experiments, simulations, and design solutions. *Transportation Science, 39*(1), 1–24.
Hirschfield, A., & Bowers, K. (Eds.). (2001). *Mapping and analysing crime data: Lessons from research and practice*. London: Taylor and Francis.
ITRC. (2011) UK infrastructure transitions research consortium. http://www.itrc.org.uk. Accessed 16 Aug 2011.
Jin, J. (2009). *Microsimulation and water demand forecasting*. PhD thesis, School of Geography, University of Leeds, Leeds, UK (available from University of Leeds, library).
Joint Information Services Committee. (2011). *Exploiting geo-spatial datasets to enhance crime analysis and related research methods*. JISC Information Environment Programme 2009–2011. http://www.jisc.ac.uk/whatwedo/programmes/inf11/jiscGEO/geocrimedata.aspx. Accessed 16 Aug 2011.
Jordan, R. (2011). *Large-number individual-level modelling of society: Regeneration and the UK housing market*. PhD thesis, University of Leeds, Leeds, UK (available from University of Leeds, library).
Jordan, R., Birkin, M., & Evans, A. (2011). Agent-based simulation modelling of housing choice and urban regeneration policy. In T. Bosse, A. Geller, & C. Jonker (Eds.), *Multi-agent-based simulation XI* (pp. 152–166). Berlin: Springer.
Lambert, P., & Birkin, M. (2012). Occupation, education and social inequalities: a case study linking survey data sources to an urban microsimulation analysis. In F. Pagliara, D. Simmonds, M. de Bok, A. Wilson (Eds.), *Advances in Employment Modelling*, Springer.
Liu, S., Chen, H., Birkin, M., Mao, B., & Guo, J. (2009, March). *Assessment of the transport policy effect on CO2 emissions using a system dynamic model*. Air Quality 2009, Istanbul, Turkey.
Malleson, N. (2010). *Agent-based modelling of burglary*. PhD thesis, University of Leeds, Leeds, UK (available from University of Leeds, library).
Malleson, N., & Birkin, M. (2011). Towards victim-oriented crime modelling in a social science e-infrastructure. *Philosophical Transactions of the Royal Society A, 369*, 3353–3371.
Morrison, R. (2007). Dynacan: Longitudinal dynamic microsimulation model. In A. Gupta & A. Harding (Eds.), *Modelling our future: Population ageing, health and aged care*. Amsterdam: Elsevier.

Nagel, K., Wolf, D. E., Wagner, P., & Simon, P. (1998). Two-lane traffic rules for cellular automata: a systematic approach. *Physical Review E, 58*(2), 1425–1437.

Office for National Statistics. (2010). *2008-based sub-national population projections for England*. National Statistics Centre for Demography. http://www.statistics.gov.uk/statbase/product.asp?vlnk=997. Accessed 15 Aug 2011.

Procter, K., Clarke, G., Ransley, J., & Cade, J. (2008). Micro-level analysis of childhood obesity, diet, physical activity, residential socio-economic and social capital variables: Where are the obesogenic environments in Leeds? *Area, 40*(3), 323–340.

Smith, D. M., Clarke, G. P., Ransley, J., & Cade, J. (2006). Food access & health: A microsimulation framework for analysis. *Studies in Regional Science, 35*(4), 909–927.

Taylor, M., Brice, J., Buck, N., & Prentice-Lane, E. (2005). *British household panel survey user manual (vol. B)*. Colchester: University of Essex.

Thompson, C., Stillwell, J., Clarke, M., & Bradbrook, C. (2010a) Understanding and validating Acxiom's research opinion poll data (Working Paper 10/06). Leeds: School of Geography, University of Leeds. Available online at http://www.geog.leeds.ac.uk/research/wpapers. Accessed 16 Aug 2011.

Thompson, C., Birkin, M., Hodgson, S., & Maclaughlin, F. (2010b, April). *The impact of target hardening policy on spatial patterns of urban crime in Leeds*. Geographical Information Systems Research in the United Kingdom (GISRUK), University College London, London.

Tomintz, M., Clarke, G., & Rigby, J. (2008). The geography of smoking in Leeds: Estimating individual smoking rates and the implications for the location for stop smoking services. *Area, 40*(3), 341–353.

Vickers, D. (2007, March 22) *Open geodemographics: The creation of the Office for National Statistics Output Area Classification*. Geodemographics and the Social Sciences Training Workshop, University of Sheffield, Sheffield, UK.

Warner, G. C., Blum, J. M., Jones, S. B., Lambert, P. S., Turner, K. J., Tan, L., Dawson, A. S., & Bell, D. (2010). A social science data-fusion tool and the data management through e-social science (DAMES) infrastructure. *Philosophical Transactions of the Royal Society A, 368*(1925), 3859–3873.

Williamson, P., Clarke, G., & McDonald, A. (1996). Estimating small area demands for water with the use of microsimulation. In G. Clarke (Ed.), *Microsimulation for urban and regional policy analysis* (pp. 117–148). London: Pion.

Williamson, P., Birkin, M., & Rees, P. (1998). The estimation of population microdata using data from small area statistics and samples of anonymised records. *Environment and Planning A, 30*, 785–816.

Wohland, P., Rees, P., Boden, P., Norman, P., Jasinska, M. (2010). *Ethnic population projections for the UK and local areas 2001–2051* (Working Paper 10/02). Leeds: School of Geography, University of Leeds. Available online at http://www.geog.leeds.ac.uk/research/wpapers. Accessed 16 Aug 2011.

Wu, B., Birkin, M., & Rees, P. (2008). A spatial microsimulation model with student agents. *Computers Environment and Urban Systems, 32*, 440–453.

Wu, B., Birkin, M., & Rees, P. (2011). A dynamic MSM with agent elements for spatial demographic forecasting. *Social Science Computer Review, 29*, 145–160.

Part IV
Validation of Spatial Microsimulation Models and Conclusion

Chapter 15
Validation of Spatial Microsimulation Models

Kimberley L. Edwards and Robert Tanton

15.1 Introduction

Spatial microsimulation models, both static and dynamic, are a useful way to estimate area-level data, whether these data are regarding health, socio-economic status or income/finance. However, in order for planners and government to be able to use and rely on these data, it is essential that the modellers can show that the estimates are an accurate representation of the real world and are reliable.

Generally, to verify the integrity of any model, it is necessary to validate the model outputs (Ballas and Clarke 2001; Oketch and Carrick 2005), using both internal and external validation methods (Edwards et al. 2010). Internal validation is the process whereby the variables that were used in the estimation of the output data are compared, so the input dataset is compared with the output dataset for those variables. This process examines the data from which the simulated dataset is drawn. External validation is the process whereby the variables that are being estimated are compared to data from another source, external to the estimation process, so the output dataset is compared with another known dataset for those variables.

Importantly, for the external validation process, the known dataset should come from a different source of data than that used in the model, that is, data not used in the original simulation (Caldwell and Keister 1996). Thus, this process examines whether the variables that have been estimated can be generalized to the population in question by comparing to an alternative dataset. That is, are the simulated data an accurate presentation of the population in question?

K.L. Edwards (✉)
School of Clinical Sciences, University of Nottingham, Nottingham, UK
e-mail: Kimberley.edwards@nottingham.ac.uk

R. Tanton
NATSEM, University of Canberra, Canberra, Australia

R. Tanton and K.L. Edwards (eds.), *Spatial Microsimulation: A Reference Guide for Users*, Understanding Population Trends and Processes 6, DOI 10.1007/978-94-007-4623-7_15,

For example, in Chap. 5, sex, age and deprivation score were used to estimate obesity. Thus, the variables sex, age and deprivation should be internally validated: compare the input dataset for these variables to the corresponding simulated dataset. Similarly, obesity rates should be externally validated: the obesity simulated dataset should be compared to a known dataset from another source, such as actual measured obesity rates in that population.

In the specific case of spatial microsimulation models validation is a massive challenge. This is because, generally, these models are used to estimate data that does not otherwise exist, perhaps due to confidentiality reasons (e.g. income or medical data for individuals) and/or because it would be expensive and time consuming to try to collect a large sample of data for the population in question (particularly as, in many countries, national sample datasets already exist, thus it would also be a duplication of both time and money). In many countries, sample surveys are conducted that provide estimates for large areas, but collecting enough sample to derive estimates for small areas is prohibitive in terms of cost and respondent burden. A national census can provide excellent small-area data which can be used for validation (and this is one of the main sources of validation data for the Australian spatial microsimulation model, SpatialMSM – see Chap. 6), but these are usually conducted every 5 or 10 years and collect a limited range of information compared to surveys.

This means often it is not possible to compare the estimates from spatial microsimulation models to actual small-area data directly. Further, with dynamic spatial microsimulation models that project into the future, there is the additional problem of validating the projections – any future scenarios will be uncertain.

15.2 Methods of Validation for Spatial Microsimulation Models

Statistical model validation is a key (arguably 'the' key) part of the process of model building and historically one that has been overlooked in the spatial microsimulation literature, whether due to non-action or non-explanation by the author. Some models automatically undertake validation as part of the simulation process, but many do not. This necessitates that the modeller spends time validating their outputs.

There is no one 'right' accepted method to validate a spatial microsimulation model. This in itself may have led to the confusion over whether to report validation and what information to report. This book has gone some way towards remedying this omission. Each methodological chapter suggests a process to assess the error in the simulated data and to validate that model. These methodologies and the different internal and external validation approaches are summarized and assessed below.

15.2.1 Validation Methodologies

In validating a spatial microsimulation model, it is firstly necessary to ensure the data are at the same spatial scale, aggregating individual level data to levels at which observed (e.g. census) datasets exist, if necessary. Thus, for internal validation, it would be necessary to aggregate the individual level data for the constraint (benchmark) variables to the observed dataset level. Likewise, for external validation, we will need to aggregate data as described in Sect. 15.2.2. There are then a number of different statistical tests that can be used to validate the data.

Some authors use TAE (total absolute error) and/or SAE (standardized absolute error) to determine the quality of the simulation (e.g. Chap. 4; Ballas et al. 1999, 2006). This method provides information on the size of the difference between the simulated and actual datasets but does not evaluate this difference. For example, a TAE of 10 might be high if the population was only 20 people but low if the population was 20,000. The SAE addresses this issue by using the population size as the denominator, but generally, authors seem to use total population, rather than the population for that categorization of the variable, which may be deemed to understate the size of the errors. Also, this measure does not provide any information on whether any differences are statistically significant. Examples of the use of SAE include early work using the SpatialMSM model (see Chin et al. 2005).

It is also possible to use regression analyses (see Chaps. 4 and 5). This technique compares the simulated data to the actual data in order to understand the fit of the simulation. Obviously, this requires an awareness of the data type, and so whether linear, poisson or logistic regression is most appropriate. To do a simple linear scatter plot, converting the data to percentages can be a useful technique (Ballas et al. 2005; Edwards and Clarke 2009). These results can be presented as scatter plots, with the simulated proportion on the x-axis and the actual (census) data on the y-axis with a trend line drawn through the datapoints. The R-squared statistic (the coefficient of determination) (R^2) is an indicator that ranges in value from 0 to 1, and it reveals how closely the simulated values for the regression trend line fit the actual (census) data. A trend line is most reliable when its R^2 is at or close to 1. Thus, with this technique, we would expect to see a high coefficient of determination for the constraint/benchmark variables (i.e. variables used in the input dataset for the model) and, if known data is available, for the variable(s) being estimated.

However, regression analysis does not give any information about the fit of the simulated data to the 'ideal' line (i.e. where $y=x$ and the simulated data is the same as the actual data). Rather, R^2 expresses the fit of the data to the 'best fit' line through that data. That is, the coefficient of determination is providing information about precision, not accuracy. Thus, a high R^2 value does not guarantee that the fit of the model to the data is good. Instead, a dummy regression line (identity line) can be drawn on the scatter plot to denote the 'perfect fit' of $y=x$. Thus, if the data had simulated perfectly, then the actual data would equal the simulated data for each small area ($y=x$), and all the points would lie on a straight line of gradient 1. From this line, a 'standard error around identity' can be calculated as

$$SEI = 1 - \frac{\sum (y_{est} - y_{rel})^2}{\sum (y_{rel} - \bar{y}_{rel})^2}$$

where SEI is the standard error around identity, y_{est} are the estimated values for each area, y_{rel} are the reliable estimates for each area from a census or other data source and y_{rel} is the mean estimate for all areas where reliable data are available. This estimate has been used in validation by both Ballas and Tanton (see Ballas et al. 2007; Tanton et al. 2011) and Chap. 6.

To more accurately address the question of model fit, the residuals of the data should be examined. These are the differences between the observed and predicted (simulated) values for each variable, which can be examined graphically (e.g. with a scatter plot). If the model is a good fit, then the residuals would behave randomly; conversely, if any non-random structure manifests in the residuals, this clearly suggests a poor model fit.

An additional statistical test is required to establish whether there are any statistically significant differences between the synthetic and real populations. One way to do this is with a *t* test (see Chap. 5). With a spatial microsimulation model validation, the data are paired (given we are comparing simulated with actual data), thus an equal variance 2-tailed *t* test can be used to determine if there is any significant difference between the two datasets (i.e. simulated and actual). Thus, if the simulation is robust, we would expect to see no significant differences between the simulated and actual values for the input variables (and estimated/output variables, if known data are available). This enables the model accuracy to be assessed, as opposed to simply its precision.

Similarly a *z*-score can be utilized for validation purposes (Chap. 8; Hynes et al. 2009; Rahman et al. 2010). This test is analogous with the *t* test described above, depending on whether population parameters are known or not. A *z*-score is a measure of how many standard deviations an observation is from the mean for those data (note, this technique assumes data are normally distributed). It is calculated by determining the difference between each individual raw score and the mean value for the population and then dividing this difference by the standard deviation for the population for each variable category. A modification is added to adjust for areas with zero counts. Calculating the *z*-score provides a value akin to a chi-squared statistic, and if this figure exceeds the critical value (1.96), the dataset is a poor fit.

An alternative method is to use a measure of accuracy (such as in Chaps. 6 and 9), which also seeks to calculate a statistical measure of the accuracy of the simulated versus actual data. It ascertains the sum of the absolute error for all constraints (benchmarks) in each area and determines whether this is larger or smaller than the population for the area. If larger, the area is flagged as being an inaccurate simulation.

15.2.2 Additional Considerations for External Validation of the Model Output

For external validation, as explained above, the process can often be hampered by a lack of known data, particularly at the appropriate (micro) spatial scale. Three methods are suggested below to circumvent this difficulty.

It may be possible to aggregate the resulting simulated data to a coarser geographic resolution in order to be able to compare the simulated data to existing 'real' survey data. This is because once a list of individuals and their attributes have been simulated, the individuals can be aggregated up to new, coarser, geographic scales at which observed datasets exist. This methodology was used in Chaps. 5, 7 and 8. This technique is useful for checking that the figures are broadly correct (i.e. at a coarse geography) (Rephann 2001) but cannot assess the correctness of the spatial distribution at the small-area level. This method was also used for the Australian SpatialMSM model described in Chap. 6 and in Tanton et al. 2011.

However, this may not always be possible, for example, if the study area simulated is very small or if it only covers part of the coarser geography that is available. For example, data simulated at the small-area level for, say, the city of Leeds in Yorkshire, could not be aggregated to the Yorkshire and Humber 'Government Office Region' (GOR) as the GOR is bigger than the city. However, if data had been simulated for the whole of the north of England, then it would be possible to use the data for the synthetic individuals that 'lived' in those micro-areas that fell within the boundaries of the Yorkshire and Humber GOR and compare the simulated data with the actual data. There may be a problem with tautology with this methodology. That is, if the survey was used in the simulation and the comparison is technically an internal one as it is being made against a dataset used in the simulation (albeit at a different spatial scale).

An important point to be aware of here is the ecological fallacy. This is where it is erroneously assumed that individuals within a group have the same characteristics as the average for that group. Thus, data that accurately describe the characteristics of a group do not necessarily apply to individuals within that group. A stereotype would be a classic example of this. For example, if a particular group is deemed to be, on average, shorter than the general population, it does not mean that every individual in that group is shorter than the general population. However, if the output from a spatial microsimulation model is an accurate representation of the individuals in that group, then their mean characteristics should concur with the real group characteristics, thus avoiding this issue.

Rather than grossing up simulated micro-level data to the coarser geography of an existing survey dataset, an alternative method of obtaining data for an external validation would be to compare simulated data to a sample of actual data for that variable, that is, to go out and collect (assuming the data do not already exist) data for one or more of the micro-areas in your simulation. There are obvious cost, time and ethical implications involved in this methodology, but it could be usefully used. For example, if obesity had been estimated for all age groups for a region, and data were routinely collected for, say, 11-year-olds in that region, then the estimates for that age group could be compared to the routine survey data and validated using the methods described above. Thus, if internal validation suggested that age was robustly simulated, and external validation showed that obesity in 11-year-olds was an accurate estimation, it would be reasonable to assume that obesity micro-level estimates for other age groups truly represented the real world.

A third approach may be to compare the simulated data to a different but correlated variable for which known data at the micro-level exist. Edwards et al. (2010) have demonstrated this successfully for obesity micro-level estimates using data for cancers known to be associated with obesity. This method relies on the data demonstrating a high ($r > 0.50$) correlation between the two variables (in this case cancer and obesity prevalence) as there is not a linear relationship between effect size and predictive power. Lower correlation figures have almost no value in prediction. It is only if the two variables are highly correlated that the predictive values become useful. Using this correlated dataset, the techniques described above can be utilized to examine the errors and whether any statistically significant differences exist between the simulated and known datasets. Accordingly, in Chap. 4, Anderson describes the use of a correlated variable (the Welsh Index of Deprivation 2005) to externally validate the simulated variable (the percentage of households below average income) due to a lack of availability of small-area data for the simulated variable. Similarly in Chap. 6, Tanton et al. use a different definition of poverty to externally assess the spatial distribution of their poverty results. This method uses a graph that compares the modelled data to the similar data, with a 45 ° line through the graph. The dispersion of the points around this line is called the standard error around identity and has been described in the previous section.

The final method of validation is to run the model for a larger area and then test the results against reliable estimates for the larger area from another dataset, for example, a sample survey. This method is a test of the method being used; if it is working for a large area, then there is some confidence that it will work for small areas. However, this method cannot replace validation of the output for small areas, as it does not capture the distribution of the estimates between the small areas; all it can test is that the method is working reasonably. This method was used, in conjunction with the SEI described above and an aggregation method described above, in Tanton et al. 2011.

15.2.3 Validation of Dynamic Spatial Microsimulation Models

As indicated earlier, there are additional complications with validating dynamic spatial microsimulation models. They are more detailed models, with more data combinations, and thus rarely are suitable micro-data available to use to externally validate. They also derive projections, forecasting expected scenarios based on assumptions about future events.

A common technique to address the problem of validating these models is to aggregate data to levels at which appropriately detailed data do exist, thus aligning the geographic resolution. The example given in Chap. 14 was to sum data to government demographic projection spatial levels, thereby also addressing the issue of assessing the validation of both current and future simulated data. The Moses model described in Chap. 11 also uses this technique for validation purposes. This capacity to match the micro-level estimates to macro-level estimates has been a key element to validation in spatial microsimulation modelling and, as described above, provides an important indication about the high-level functioning of the model. Thus, to take this to the next stage, authors often adjust the simulated data as necessary (O'Donoghue 2001).

However, this method simply serves to illustrate that the dynamic spatial microsimulation model is a good model to use to disaggregate government macro-level forecasts of society, thus relying on the quality of the underlying forecasts in the first place. That is, if the forecast is poor, then whether or not the validation method shows a good fit is not representative of whether the micro-level simulations are truly representative of the real world or not (i.e. a poor fit to a poor estimate tells us nothing). Better methods perhaps and solutions to these problems, as suggested in Chap. 14, may be to use backcasting methods (Birkin and Malleson 2010) or to model only specified scenarios which allow for different futures to be examined (Wohland et al. 2010; De Beer et al. 2010).

15.3 Reasons Why Validation May Be Poor

Exasperatingly, there is both art and science to spatial microsimulation modelling. High errors and poor validation could be due to a number of reasons.

It is important that the sample population is sufficiently large (Huang and Williamson 2001). If too small, there will likely be too few individuals at the extremes of the spectrum, for example, the very poor or wealthy, those with rare combinations of characteristics such as same-sex partnerships or ethnic minorities. Thus, the precision and accuracy of the model may tail off at these extremes and for those areas with relatively high proportions of these individuals. Thus large, heterogeneous population datasets are likely to be associated with lower errors. It is worth considering combining survey datasets in order to achieve this.

Little research has been undertaken to assess whether few (Edwards and Clarke 2009) or many (Tanton and Vidyattama 2010) constraint variables produce a better fitting model. It will likely be different for different research questions and different underlying datasets, varying with, for example, the strength of the correlations between constraint and output variables in the datasets. Also more constraints, and categories of constraint variables, lead back to the problems at the extremes of the populations; do any individuals in the survey match these specific combinations? If not, small number problems can be encountered. What is clear is that different constraint combinations will lead to different output datasets (Huang and Williamson 2001), thus different validation results.

Large variations between simulated and actual data, resulting in large errors and poor validation results, could be caused by legitimate regional differences, that is, spatial attributes that are not included in the model (Birkin and Clarke 2011). This is particularly relevant for spatial microsimulation models that only use socio-economic variables to determine the simulation output.

Of course, poor validation could also indicate that the model is a poor fit and the assumptions underlying the simulation, such as choice of datasets and/or constraint variables, need to be revisited and revised in order to improve the fit. To proceed to use a model that does not fit the data well would, in turn, mean that the questions the model is being used to investigate (whether for government policy or scientific investigation) cannot be provided with good, or reliable, answers.

15.4 Conclusions

This chapter has explained the common methodologies for validating spatial microsimulation models as well as, hopefully, enforcing the point that this is an essential step in the modelling process and one that cannot be skipped.

Research is currently lacking in comparing and assessing these different methodologies to determine which is 'best' or the gold standard. However, given the choice of technique is largely driven by what data are available, and that this is a primary difficulty for these models, it is likely that authors would still be driven by the availability of data. It may also be that some methods are better for particular spatial microsimulation algorithms but not for others.

Importantly, researchers should be transparent about their validation methodologies and realistic about the strength of the estimates and accuracy of the models. Further, collaborating with colleagues from other disciplines (such as statistics, economics, medicine) and pooling resources and knowledge would facilitate improvements in validation techniques. Also, continuing to tackle real-world problems with these models should be encouraged (Wu et al. 2008; Gilbert 2009). Working with local authorities and government, explaining the methods and the rationale in lay language, will lead to increased knowledge about, and acceptability of, these models. In some way, while it has little statistical rigour, the authors have found that one of the best ways of validating these models is to show maps of the results to practitioners or people in the field, and they will quickly tell you if the results look reasonable or not. This also assists with increasing the impact of these models in addressing important social, medical and economic questions.

References

Ballas, D., & Clarke, G. (2001). Modelling the local impacts of national social policies: A spatial microsimulation approach. *Environment and Planning C: Government and Policy, 19*, 587–606.

Ballas, D., Clarke, G., & Turton, I. (1999, July). *Exploring microsimulation methodologies for the estimation of household attributes*. Paper presented at the 4th International Conference on GeoComputation, Mary Washington College, Fredericksburg, VA.

Ballas, D., Rossiter, D., Thomas, B., Clarke, G., & Dorling, D. (2005). *Geography matters: Simulating the local impacts of national social policies*. York: Joseph Rowntree Foundation.

Ballas, D., Dorling, D., Anderson, B., & Stoneman, P. (2006). *Assessing the feasibility of producing small area income estimates: Phase I project report*. Sheffield: Department of Geography, University of Sheffield.

Ballas, D., Clarke, G., Dorling, D., & Rossiter, D. (2007). Using SimBritain to model the geographical impact of national government policies. *Geographical Analysis, 39*(1), 44–77.

Birkin, M., & Clarke, M. (2011). Spatial microsimulation models: A review and a glimpse into the future. In J. Stillwell & M. Clarke (Eds.), *Population dynamics and projection methods*. London: Springer.

Birkin, M., & Malleson, N. (2010, November 10–13). *An investigation of the robustness of a dynamic microsimulation model of urban neighbourhood dynamics*. Paper presented to the North American Regional Science Council (NARSC), Denver, CO. http://www.geog.leeds.ac.uk/fileadmin/downloads/school/people/academic/m.birkin/backsim_paper_v2.pdf. Accessed Sept 2011.

Caldwell, S., & Keister, L. (1996). Wealth in America: Family stock ownership and accumulation, 1960–1995. In G. P. Clarke (Ed.), *Microsimulation for urban and regional policy analysis* (pp. 88–116). London: Pion.

Chin, S. F., Harding, A., Lloyd, R., McNamara, J., Phillips, B., & Vu, Q. (2005). Spatial microsimulation using synthetic small-area estimates of income, tax and social security benefits. *Australian Journal of Regional Studies, 11*(3), 303–335.

De Beer, J., Van der Gaag, N., Van der Erf, R., Bauer, R., Fassmann, H., Kupiszewska, D., Kupiszewski, M., Rees, P., Boden, P., Dennett, A., JasiDska, M., Stillwell, J., Wohland, P., De Jong, A., Ter Veer, M., Roto, J., Van Well, L., Heins, F., Bonifazi, C., & Gesano, G. (2010). *DEMIFER Demographic and migratory flows affecting european regions and cities* (Applied Research 2013/1/3, Final Report | Version 30/09/2010). Luxembourg: ESPON.

Edwards, K. L., & Clarke, G. P. (2009). The design and validation of a spatial microsimulation model of obesogenic environments in Leeds: SimObesity. *Social Science and Medicine, 69*, 1127–1134.

Edwards, K. L., Clarke, G. P., Thomas, J., & Forman, D. (2010). Internal and external validation of spatial microsimulation models: Small area estimates of adult obesity. *Applied Spatial Analyses and Policy*. doi:10.1007/s12061-010-9056-2.

Gilbert, N. (2009). *Microsimulation and agent-based modelling* (Plenary address). Ottawa: International Microsimulation Association.

Huang, Z., & Williamson, P. (2001). *A comparison of synthetic reconstruction and combinatorial optimisation approaches to the creation of small-area microdata* (Working Paper 2001/02). Department of Geography, University of Liverpool [online]. http://pcwww.liv.ac.uk/~william/microdata/workingpapers/hw_wp_2001_2.pdf. Accessed Sept 2011.

Hynes, S., Morrissey, K., O'Donoghue, C., & Clarke, G. (2009). Building a static farm level spatial microsimulation model for rural development and agricultural policy analysis in Ireland. *International Journal of Environmental Technology and Management, 8*, 282–299.

O'Donoghue, C. (2001). Dynamic microsimulation: A methodological survey. *Brazilian Electronic Journal of Economics, 4*. http://www.microsimulation.org/IMA/BEJE/BEJE_4_2_2.pdf; https://www.vengroup.com/ve-net/Library.nsf/ab283684d03f231d80256b520047d321/426E6F6D3E95DF4880256F6A0044E430/$file/DynamicMicrosimulation.AMethodologicalSurvey.pdf. Accessed Sept 2011.

Oketch, T., & Carrick, M. (2005, January). *Calibration and validation of a micro-simulation model in network analysis*. Proceedings of the 84th TRB Annual Meeting, Washington, DC.

Rahman, A., Harding, A., Tanton, R., & Liu, S. (2010, July/August). *Simulating the characteristics of populations at the small area level: New validation techniques for a spatial microsimulation model in Australia*. Proceedings of the Social Statistics Section of the American Statistical Association Conference, Vancouver, Canada, pp. 2022–2036.

Rephann, T. (2001). *Economic-demographic effects of immigration: Results from a dynamic, spatial microsimulation model*. The 2001 Annual Meeting of the Mid-Atlantic Division of the Association of American Geographers, College Park, MD. http://www.equotient.net/papers/immmic.pdf, Accessed Sept 2011.

Tanton, R., & Vidyattama, Y. (2010). Pushing it to the edge: Extending generalised regression as a spatial microsimulation method. *International Journal of Microsimulation, 3*(2), 23–33.

Tanton, R., Vidyattama, Y., Nepal, B., & McNamara, J. (2011). Small area estimation using a reweighting algorithm. *Journal of the Royal Statistical Society: Series A (Statistics in Society)*. doi:10.1111/j.1467-985X.2011.00690.x.

Wohland, P., Rees, P., Boden, P., Norman, P., & Jasinska, M. (2010) *Ethnic population projections for the UK and local areas 2001–2051* (Working Paper 10/02). Leeds: School of Geography, University of Leeds. http://www.geog.leeds.ac.uk/research/wpapers. Accessed Sept 2011.

Wu, B., Birkin, M., & Rees, P. (2008). A spatial microsimulation model with student agents. *Computers Environment and Urban Systems, 32*, 440–453.

Chapter 16
Conclusions and Future Research Directions

Graham Clarke and Ann Harding

16.1 Background

While microsimulation has established itself across the industrialised world as a useful tool for estimating the distributional impacts of policy change upon households and individuals, such models have typically provided results at the *national* level. Policy makers, however, have been keen to understand more about the characteristics of *small-area* populations and to assess the small-area impact of changes in policy or demographics. Over the past two decades, researchers have responded to this challenge.

The speed of development has gathered pace since the 1987 journal article on spatial microsimulation by Clarke and Holm (1987). The edited volumes arising from the international meetings of microsimulators testify to the growth in spatial microsimulation. For example, none of the chapters in the book arising from the 1993 microsimulation conference (Harding 1996) or from the 1997 conference (Gupta and Kapur 2000) covered spatial microsimulation. Mitton et al. (2000) covered the 1998 conference and contained one chapter out of 14 on the spatial SVERIGE model. After 2000, the spatial microsimulation field evolved rapidly. By 2007, the two books arising from the 2003 conference contained descriptions of four spatial models (out of 22 model descriptions) and one chapter with a spatial microsimulation application (Gupta and Harding 2007). By 2009, however, spatial microsimulation had developed so rapidly that the first four chapters in the Zaidi et al. book (which covered the 2007 conference) were spatial microsimulation applications.

Some earlier approaches to small-area estimation came out of the discipline of statistics, as statisticians attempted to link survey outcome or response variables to

G. Clarke (✉)
School of Geography, University of Leeds, Leeds, UK
e-mail: g.p.clarke@leeds.ac.uk

A. Harding
NATSEM, University of Canberra, Canberra, Australia

R. Tanton and K.L. Edwards (eds.), *Spatial Microsimulation: A Reference Guide for Users*, Understanding Population Trends and Processes 6, DOI 10.1007/978-94-007-4623-7_16,

a set of predictor variables known for small areas, in an attempt to 'borrow strength' from other data (Rao 2002, 2003; Rahman et al. 2010). In the United Kingdom, alternative approaches to the creation of synthetic small-area microdata were developed by geographers, such as Birkin and Clarke (1989), who used a technique called synthetic estimation to attach conditional probabilities to each individual (e.g. to estimate income). Williamson pushed the boundaries of the emerging discipline, using combinatorial optimisation to reweight survey data and to develop methods of validation and goodness of fit (Williamson 2001; Voas and Williamson 2000). The field of spatial microsimulation in the UK has since blossomed, expanding to geographers at other universities and institutes and being utilised in an ever-growing number of applications (with the UK experience being well documented in this volume and summarised in the next section of this chapter).

In contrast, the Australian approach has emerged from economists and public policy experts, almost exclusively those located at the National Centre for Social and Economic Modelling (NATSEM) at the University of Canberra. NATSEM's staff originally specialised in *national*-level microsimulation models, which were used for a diverse range of policy analyses, spanning taxes, government benefits, education and so on. One of the frustrations associated with the sample survey data that underlaid many of the NATSEM models in the 1990s was the lack of geographic detail (which was required to maintain the confidentiality of the respondents to national sample surveys conducted by the Australian Bureau of Statistics (ABS)). To overcome this deficiency, NATSEM staff began to experiment with the production of synthetic small-area microdata, with a paper on the geodemographics of the aged being one of the first outputs from this new stream of research (Harding et al. 1999).[1]

In the intervening 13 years, the spatial microsimulation techniques employed by NATSEM have improved substantially (see Tanton et al. 2011). The reweighting software used has changed significantly; greater experience has resulted in new techniques to assist in selection of the most relevant census benchmarks for the purpose in hand; there is a greater understanding of the number of census variables that the sample survey can be successfully matched to and the number of applications has grown substantially. (Detailed descriptions of the spatial microsimulation model construction can be found in Chin et al. (2005) for the earlier versions and Cassells et al. (2010) for the more recent models.) Relevant applications have included studies of neighbourhood poverty and inequality (Harding et al. 2006, 2011a; Tanton et al. 2009a, 2010; Miranti et al. 2011; Gong et al. 2011), predicting the need for aged care services and regional disability estimates (Lymer et al. 2006, 2008a, b) and superannuation savings at the small-area level (Vidyattama et al. 2011).

A further innovation is that NATSEM's static microsimulation model, STINMOD, has been successfully grafted onto the synthetic individual and household small-area microdata. STINMOD models the 'morning after' impact of changes in income tax,

[1] The authors would like to thank and acknowledge the NATSEM staff who contributed to the early days of spatial microsimulation at NATSEM, including Otto Hellwig, Anthony King, Tony Melhuish, Susan Day and Elizabeth Taylor.

tax concessions and rebates, social security system and government cash payments to families with children (Lloyd 2007). This innovation means that research using the new synthetic microdata is not restricted to variables on the sample survey that was matched to the census data. Examples utilising these path-breaking techniques include local estimates of disposable income, income tax and social security benefits (Chin et al. 2005) and of housing assistance (McNamara et al. 2007). Further, STINMOD can be used to model the impact of *policy changes* in taxes and benefits (such as the impact on effective tax rates of a liberalisation of the income-tested family payment (Harding et al. 2009), raising the age pension for singles (Tanton et al.. 2009a), and the neighbourhood impact of the Australian government's stimulus package following the global financial crises (Tanton and Vu 2010)).

Finally, in Australia and the UK, attempts are being made to create forecasts for small areas. In Australia, this has included the future need for aged care and child care services (Harding et al. 2011) and methodologies for projecting small-area statistics (Vidyattama and Tanton 2010).

16.2 Chapter Summaries

As noted above, the chapters in this book are testimony to the fact that the field of spatial microsimulation has progressed rapidly over the last two decades. From humble beginnings in a small number of research centres (largely driven by early theoretical explorations), the field has mushroomed to include all parts of the world and a huge variety of application areas. This book provides a good illustration of that fact: authors based in the UK, Australia, Sweden and Ireland and application areas involving a broad range of socio-environmental issues. In addition, there are a number of chapters which report on progress with major theoretical developments, especially in relation to data matching/estimation techniques and the traditional thorny issues of model calibration/validation. For us, the book makes four major contributions. First, it provides a unique guide to implementing a spatial microsimulation project. Those who are just beginning to become interested in the techniques will surely welcome the step by step guides provided in the early chapters. Second, it discusses alternative techniques for estimating microdata when such data are not available from published sources. Third, it provides a detailed assessment of various calibration and validation techniques. Fourth, it shows the value of microsimulation through a number of case studies of important application areas around the world. We shall discuss each of these issues as they appear in each chapter.

To aid understanding of how the various chapters contribute to knowledge about spatial microsimulation, Fig. 16.1 shows the processes typically involved in construction of a spatial microsimulation model. The first step is to determine output requirements for the study being undertaken and select relevant benchmarks from the census small-area tables. For example, a study of obesity might require selection of different benchmarks to one on poverty rates. The second step is to ensure that, as far as possible, the variables on the sample survey and the census data are defined in the same way and that the scope is comparable. Typically, the available census

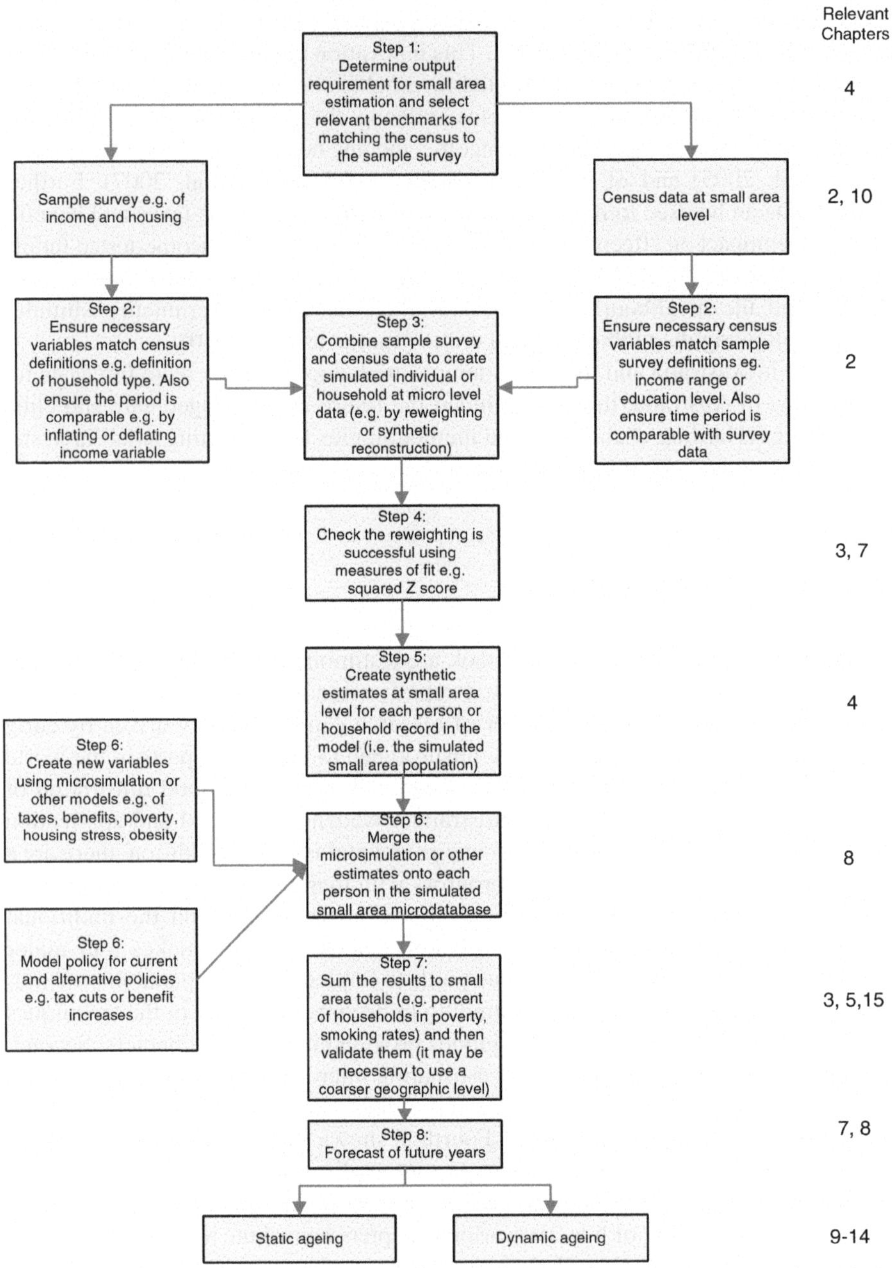

Fig. 16.1 Overview of steps required to construct a spatial microsimulation model

data will be from a different time period to the sample survey data so that uprating or deflation of incomes, housing costs, etc., is required for either the census or the sample survey before they can be matched.

The third step is to match the sample survey unit records to the census benchmarks so as to create synthetic spatial microdata. The techniques for undertaking this step vary, as the chapters in this volume make clear. The fourth step is to check that the reweighting has worked and that satisfactory and sensible synthetic population estimates have been created. The fifth step involves the summing of the individual household or person results so as to create small-area estimates for the variables under analysis. Many researchers stop at this point if their sole intent is to create synthetic small-area populations and/or to analyse existing variables within the small-area data arising from the sample survey used in the reweighting (such as housing tenure).

However, many other researchers want to create *new variables* that are not present in either the census or the sample survey. Thus, the sixth step for those who have linked their spatial microdatabase to a microsimulation model or an econometric or other model is to create these additional variables and then merge them onto the microdatabase. This step may also include variables demonstrating the impact of policy change (e.g. the researcher might create new variables such as 'income tax paid under current system' and 'income tax paid after tax cut'). The seventh step is then necessary for those researchers adding these new variables, as a further round of validation is required here, to ensure that the results appear reliable (e.g. see Chin et al. 2005).

The eighth possible step, which is a more recent development, is to 'age' the spatial microdatabase through time to create forecasting versions of the model. Examples included in this book include 'static ageing' and 'dynamic ageing'.

As the above discussion makes clear, creating a spatial microsimulation model is a demanding exercise. There is no doubt that one major explanation of why there are not more applications of microsimulation is that the technique is hard to implement and thus costly – perhaps not in terms of understanding the concepts but in terms of learning the programming techniques. Unlike many other statistical and even mathematical modelling methods, there are few examples of the major steps that building a real model entails. Chapter 2 by Rebecca Cassells, Riyana Miranti and Ann Harding has provided a good introduction to the types of data required to build a spatial microsimulation model. This is a good starting point for a new user since the chapter discusses the principles of how data can be combined from geographical data sources such as censuses and survey data (which often have very limited spatial data). The former provides a link between household type, geodemographics and small-area geographies, whilst the latter can provide valuable information on the interdependencies between core variables (age, sex, social class, etc.) and additional variables of interest for a particular application area (health status, crime, educational attainment, etc.). Without these two basic ingredients, microsimulation cannot really work, although it may be appropriate to alternatively match two geographical data sets which combine different variables of interest.

Once the user has understood the data requirements, the next major issue is how to generate the small-area estimates. In other words, what techniques are available to undertake the matching process? Chapters 2, 3, 4, 5, 6, 7 examined various alternative ways of estimating small-area data. Paul Williamson (Chap. 3) has been a leading expert on the pros and cons of different techniques ever since a key publication with colleagues in Leeds in 1998 (Williamson et al. 1998). In this chapter, he has explored two commonly applied techniques: 'synthetic reconstruction' and 'combinatorial optimisation'. The first is a technique widely used in early applications of spatial microsimulation from the UK (i.e. Birkin and Clarke 1988, 1989). It involves estimating variables through attaching conditional probabilities – for example, given age, sex and social class and what is the probability that an individual will be a smoker? Then random numbers would typically be used to determine whether each individual is classed as a smoker or not in the model.

Combinatorial optimisation is one of a number of techniques for reweighting survey data so that it best fits small-area distributions. In effect, individuals in samples are cloned whenever they match against similar household profiles obtained from censuses. Using new calibration techniques, Williamson argues the performance of the optimisation models is superior (this involves the specification of measures of fit based around the *Z*-score and two derivations: SZ2 and RSSZ). That said, synthetic reconstruction is still a valid technique and especially useful when estimating income, which is notoriously difficult data to obtain even from surveys (see also the contrasting evaluation of these two different methods in Harland et al. 2011). Birkin and Clarke (1989) used this technique for their estimates of income in Leeds in the mid-1980s.

In Chap. 4, Ben Anderson has updated this type of approach to estimate income for Wales. In this case, 'iterative proportional fitting' was the technique used to estimate the conditional probabilities. This technique is well suited for estimating the likelihood of income from employment as well as a raft of different types of benefit. Both of these chapters also introduced the notion of constraint variables. An important choice to be made in any application is which variables to use to undertake the matching process. Sometimes this is an easy decision as there may only be a few variables common to both data sets. However, often there is more choice, and the end results can be very dependent on those choices made (see also Edwards and Clarke 2009). This is a hugely important step in the microsimulation process, and more work needs to be done in the future on the choice of constraints and the differences this may produce in the final model. Elsewhere, Smith et al. (2011) have provided a novel approach by varying the constraints chosen according to the demographics of the area being modelled. They undertook a type of cluster analysis of the geodemographics of the population of Leeds and used different constraints to match on variables more appropriate to those clusters.

The discussion in Chap. 3 by Williamson has been extended in Chap. 5 by Kimberley Edwards and Graham Clarke. They noted that the combinatorial optimisation procedure can be solved in various contrasting ways. The chapter first debated the pros and cons of probabilistic and deterministic techniques for data imputation. Having chosen the latter, the chapter discussed the data and methodology used to estimate the small-area prevalence of obesity in Leeds. The chapter also introduced two types of validation – first, against the data contained in the surveys (i.e. do the estimates fit against known totals of variables in the census or sample, so-called internal variables) and second, data held outside the surveys (so-called external variables). This chapter also provided an interesting application of microsimulation – namely, the estimation of obesity rates at the small-area level. (This is another variable which is always difficult to access due to confidentiality issues: hence the need for robust estimates if local resource targeting is to be adopted.)

Income is a key variable that was explored again in Chap. 6 by Robert Tanton, Ann Harding and Justine McNamara, this time for all statistical local areas (SLAs) in Australia. In addition, the application included the estimates of benefits and housing stress. The chapter fitted nicely into the discussion to date – exploring a reweighting method based on a generalised regression technique. The models were validated in three ways, which again show the importance of robust estimates. When estimating microdata (whether it is for households or individuals), the outcomes are in a sense estimates of missing data. Thus policy makers need to be especially sure that the estimates are robust and considerably better than 'guesstimates'. The results in this chapter were validated against SLA data, state data and proxies in the census for the income and benefit indicators being estimated. The set of maps showed the immediate impact of such research – a striking set of income maps for Australia.

Further techniques for generating small-area data sets were introduced in Chap. 7 by Niall Farrell, Karyn Morrissey and Cathal O'Donoghue. They also introduced the SMILE model for examining the socio-economic and environmental impacts of policy change in rural Ireland. This model has been in existence since 1999 (in various formats), and more details will appear in a complimentary book to this one, due for publication in 2012 (Ballas et al. 2012). The book is complimentary in the sense that it focuses on one (broad) model only and explores the range of potential applications that can be built on such a foundation. It will also concentrate exclusively on issues of policy importance to rural communities and rural service providers. The data reweighting techniques used in the chapter in this book were simulated annealing and quota sampling. The authors discussed the relationships with the techniques discussed in Chaps. 3, 4, 5, 6. The validation of the income and benefit estimates in this chapter introduced a new problem not discussed so far. When validating against known data, the estimates may not fit in certain locations due to spatial heterogeneity – that is, local or geographical factors not visible in the surveys being reweighted. For example, the estimation of car ownership rates from surveys based on population characteristics alone cannot take into account, for example, the proximity of the residents to key work locations (i.e. city centre living might in fact make car ownership less important by allowing individuals to walk to work).

The solution in this chapter was to realign the small-area estimates based on real data known at higher levels of spatial resolution (i.e. county data). This is a well-known technique in many spatial and aspatial (countrywide) microsimulation models (e.g. see Caldwell and Keister 1996). The authors conclude that 'on completion of the alignment process, SMILE offers a fully representative profile of labour force participation and market incomes at both the household and small-area level'.

Outside the contents of this book, there are other ways to try and correct for spatial heterogeneity. Birkin and Clarke (2011), for example, suggest that adding geodemographic classifications to the estimated data might allow the model builder to adjust the results accordingly. To return to the car ownership discussion introduced above, for example, estimates made for locations labelled in some sense as 'suburban' would introduce a boost in local rates estimated whilst a 'city centre location' would cause the model to reduce the estimates made.

There was a shift in focus in Chap. 8. Karyn Morrissey, Graham Clarke and Cathal O'Donoghue have shown how the SMILE model can be enhanced by adding new types of modelling frameworks. In this case, they introduced spatial interaction models (SIM) and linked them formally to the microsimulation model. SIMs are usually mesoscale models, in the sense that households or individuals are aggregated into zones and flows are modelled from these zones to service locations (shops, schools, health centres, etc.). The demand within these zones is often based on the aggregate census information for those zones. When combined with microsimulation models, these interaction models can be made more local in three ways. The first is by building the disaggregated population estimates into the zonal demand estimates. An example of this is provided by Nakaya et al. (2007), who obtained survey data relating to various consumer groups not available directly in the Japanese census. By estimating the location of these consumer groups through microsimulation, they were able to build a citywide SIM of shopping flows based on this detailed survey.

Second, SIMs can be used to estimate small-area accessibility indicators which can then be fed into the microsimulation model as an additional variable when considering access to services. Thus, in Chap. 8, Morrissey et al. added accessibility scores from the SIM to the SMILE model to see if they could better predict the long-term illness variable, which could be influenced by access to health-care services as much as individual characteristics. The third possible link, not discussed in the chapter, is to build individual SIMs for each household in the microsimulation model. Thus, each individual or household would be given additional behavioural characteristics, such as where they shop, go to work, etc. This starts to lead into a discussion of the relationship between microsimulation models and agent-based models (see Wu et al. 2011).

From Chap. 9 onwards, the focus shifted to issues relating to forecasting. Thus, instead of 'simply' estimating current socio-economic characteristics, the dynamic models attempt to 'age' the population forward and update those socio-economic characteristics to some future point in time. In Chap. 9, Yogi Vidyattama and Robert Tanton discussed a static ageing process which is in fact a short-hand approach to population dynamics. They aged the population forward at yearly intervals but ignored any behavioural changes (although they did model future income earnings and labour market characteristics). The rationale for that is the very high degree of complexity, cost and data requirements to build a fully dynamic microsimulation model. The main advantage of the approach is to allow policy makers, in this case in Australia, to disaggregate the official population projections of the state governments. Indeed, the authors noted that although the model is relatively simple in dynamic forecasting terms, the results show a very good level of correlation with official population projections, even for small areas. The standard error around identity (SEI) for the 2027 projections, using various population age groups compared with the DOHA projections (Department of Health and Ageing 2009), was as high as 95% in some cases.

In Chap. 10, Robert Tanton and Kimberley Edwards summarised some of the main difficulties with static spatial microsimulation models, which helped to focus attention again on dynamic models. In addition to the calibration issues that most authors address, they concentrated on three issues: data limitations, benchmarking against constraints and the representativeness of the data being reweighted. The first relates to variables being defined differently in the data sets to be matched. This is especially problematic when the same variables are grouped into different classifications – with age groups being a classic example. For example, if the census categorises population into age cohorts, 15–25, 26–35, etc., and the survey uses 12–18, 19–30, etc., then there are major problems to be addressed before matching can take place. The second problem relates back to the choice of the number of constraints selected for the match. The authors noted that there is a simple correlation between the use of more variables and the greater computational time and costs. There is no simple solution to this problem. As the authors noted, the user must decide on whether the error margins are acceptable with fewer constraints; if not, then more computer time must be spent on increasing the number of constraint variables. The third problem is also related to the spatial heterogeneity issues discussed above. This time, the emphasis is not on the impact of geographical factors not captured in the survey data but is the fact that a national survey may not differentiate internally between regional variations in key variables. So the question is, if simulating a population in, say, Florida in the US, is it right to use all households in the US to match to or simply those in Florida? The latter might be more accurate but gives fewer households to effectively 'clone'. Again the authors discuss some interesting alternative ways to address this problem.

In Chaps. 11, 12, 13, 14, 15, the book concentrates on dynamic models. In Chap. 11, Belinda Wu and Mark Birkin discuss MOSES, a major UK Government sponsored dynamic microsimulation model. They first discussed the major difference between static and dynamic models and provide a brief review of the history of such models. This model extends the type of ageing seen in Chap. 9 by adding a fuller set of behavioural change variables. This means that each individual is not only aged in the dynamic model but is tested each year for the probability of death, marriage, giving birth, migration and a change in health status. These probabilities are derived from known data sets relating to birth rates, death rates, etc., and a Monte Carlo type simulation is used to test whether individuals are deemed to move to a new state (get married, give birth, etc.). Again, some would argue this is not a full dynamic model, in the sense that the labour market is downplayed and other behavioural components are ignored. However, the model must be functional in the sense that it can handle the key variables that the authors were most interested in. The model is validated using other population benchmarks and forecasts. The authors concluded by addressing the policy environment. The evolution of population structures and various demographic changes can be used to drive various location-based policies. 'For instance, the ageing trends in certain suburban areas may promote changes in health service and public transport service provision in order to enable easy access to such services for the old and frail in the area'.

Chapter 12 provides a very good overview of many issues relating to the construction of dynamic models. Einar Holm and Kalle Mäkilä have drawn on years of experience of dynamic modelling. This Swedish group of researchers has been important innovators in microsimulation for many years, spurred on by the fabulously detailed microdata sets available in Sweden. They provided and discussed a very useful wish list of data and techniques required for dynamic microsimulation and made a commentary around each. In effect, they offered us a useful list of design principles which every new modeller would be well advised to visit. They also introduced some new issues to be addressed, such as debugging computer programmes. This is never an easy task, but the idea of plotting individual biographies produced from the model is an excellent suggestion for helping make sure the results are sensible and logical. This is also one of the techniques being used to validate the APPSIM model that NATSEM is constructing (Harding et al. 2010).

In Chap. 13, Dimitris Kavroudakis, Dimitris Ballas and Mark Birkin demonstrated the usefulness of a dynamic model focused on a particular application area. Many dynamic models are built to be multi-functional. In other words, there are many variables added in order that users and policy makers can pick and choose which outputs to examine. In some cases, however, needs could be more focused. In this chapter, the authors modelled educational attainment, university entrance and the inequalities produced within the UK education system. The dynamics works by tracking an individual's social class and income over time using the British Household Panel Survey (which is a longitudinal data set of immense wealth in terms of the number of variables contained within it). The likelihood of going to university is then estimated, along with subject of study and likelihood of graduation. The model can reveal hot and cold spots of university attendance and thus make a valuable contribution to the (UK) debate in higher education around widening participation, especially to lower-income students. This type of application should be an important future dimension of dynamic microsimulation. Too often model builders feel the need to include everything in dynamic models. They soon become cumbersome and difficult to programme, and their output can be difficult to extract. Thus, customised models such as this educational attainment example should be strongly encouraged.

In Chap. 14, Mark Birkin has presented an agenda for future dynamic models. Again this is a modeller with enormous experience in microsimulation (and now agent-based) models. He was a major pioneer of the synthetic reconstruction models at the University of Leeds in the 1980s (Birkin and Clarke 1988, 1989) and has also been a driver of the move towards reweighting models from the 1990s onwards (Williamson et al. 1998). He has built on the material relating to MOSES in Chap. 11 to highlight some major research issues for the future. Again he has added some useful observations to areas already highlighted above, such as the future richness of linking microsimulation to agent-based models. However, he also has outlined some interesting ideas on other issues. In terms of validation, why not try backward simulation – that is, can you get to the present state of affairs by simulating past events? What one may learn from the dynamics of past change could be invaluable in helping to build better predictive models of the future. If one cannot replicate past trends, then perhaps key variables are missing from the model. He also called for greater openness and transparency in model design. Although SimObesity in Leeds is now available as a package-type solution to starting microsimulation, more open-source code would be valuable, in order to stop the 'wheel being reinvented' in every new application. It seems we are still a long way though from a totally generic model (and there is debate about whether that is even possible). Finally, Birkin noted the importance of discussing outputs with end users at the earliest possible opportunity. Not only can they often add valuable new data sources (especially for validation), but they can offer considerable guidance on model outputs and associated performance indicators.

In the final main chapter, Kimberley Edwards and Robert Tanton concluded with a neat summary of the issues surrounding validation. This has been a key theme throughout the book of course, but it is useful to see the various ideas rounded up into a final chapter. For many end users, validation is the key to accepting model results, especially when microsimulation often aims to estimate missing data. The authors offered a range of validation techniques to help the new researcher to get to grips with validation (and see again Paul Williamson's chapter, which includes valuable indicators for validation). In addition to the discussion of these key calibration indicators, more emphasis needs to be given in the field to external validation. As more and more data sets become available in the public domain, it is getting increasingly possible to validate model outputs, at least somewhere in the world! A good illustration is the recent work on smoking rates by Smith et al. (2011). The lack of small-area data on smoking patterns has been the impetus for a number of recent models to estimate smoking rates in different countries or areas (Tomintz et al. 2008, for example). Often these have not been externally validated, given the lack of comparable small-area data. However, the availability of data in New Zealand has allowed Smith et al. to build a model that they could externally validate. The good news for microsimulation modellers everywhere is the good fit they were able to show when comparing the spatial smoking rates against their model predictions.

16.3 Future Research Directions

As the above discussion highlights, major advances have been made in the past decade in the techniques used to create and validate spatial microsimulation models: in combining microsimulation with other modelling approaches (such as agent-based modelling) and in the subject areas covered by the modelling. However, there remain many possible directions for future research.

Looking first at the *modelling techniques*, there is ample scope for further improvement. It would be desirable if the spatial microsimulation community were able to continue to analyse which of the various reweighting/synthetic reconstruction techniques is most accurate – or to identify whether one approach is superior for some applications while another approach is to be preferred for other applications. This means further research on validation of the synthetic population estimates produced by researchers. It also suggests that it is important to publish further work on areas where spatial microsimulation is *not* successful. As Ballas et al. observe, 'the geographical simulation method is not suitable for the prediction of rare or badly reported events, such as drug use' (Ballas 2005, p. 117). NATSEM encountered similar difficulties when attempting to estimate small-area domestic violence estimates, while the UK MOSES team ran into comparable problems when trying to estimate student migration into and out of areas.

Additional research is also needed on the most appropriate spatial unit to use in spatial microsimulation. Many researchers have tested results using very refined spatial levels (such as the ward in the UK and the collection district in Australia) but have found coarser geographic measures produce more reliable estimates – which is, for example, one reason why NATSEM uses the statistical local area as its geographic level.

There remains enormous scope for researchers to improve the validation of results, tackling estimates of variance, the calculation of confidence intervals and the ongoing assessment for which external data sources provide the best benchmark against which to assess the simulated results (Rahman et al. 2010). Additional research is desirable on whether and how one aligns the synthetic estimates to other benchmark data which suggest that the synthetic estimates are too high or too low. Systematic analysis of how, and when, to align model results would be of great interest.

It is also clear that there will be ongoing demand for spatial microsimulation estimates, and as confidence in the spatial techniques grows, we can expect more modellers to link their static models of taxes and benefits to synthetic small-area population microdata.

A related issue is that the subject areas to which spatial microsimulation is applicable will grow rapidly, with the recent introduction of models of diabetes providing a good example of this phenomenon. There seems little doubt that other spatial health and aged care models will be developed in the future. Other subject areas for possible future applications include analysis of consumers and the best location of shopping centres, driver behaviour, transport, service use and supply, water and electricity consumption, energy use, waste disposal and use of information and communication technologies (such as mobile phones or the Internet).

Another area to flag is the likely rapid development of other types of models to 'mix and match' with synthetic small-area microdata. This will include further exploration of the linkages between agent-based modelling and spatial interaction models with synthetic small-area microsimulation models. Finally, ongoing efforts to create forecasting versions of the spatial microdatabases will be welcome, as practitioners grapple with the issue of the 'static' versus 'dynamic' ageing issue. Simply reweighting or aligning the synthetic microestimates for a current year to comparable population groups in 20 years time is a relatively easy option compared to the alternative. Dynamic microsimulation involves updating the characteristics of each individual within the model during every time period (typically 1 year at a time). However, apart from the numerous data and computing requirements associated with dynamic microsimulation, one crucial but frequently overlooked issue is that the probabilities of events happening to individuals are usually estimated from a national sample survey, which effectively means they provide national probabilities rather than small-area probabilities. This can be very important when the characteristics of the small-area populations under review are very different to the national population averages.

As even this brief discussion indicates, there are myriad research opportunities in the expanding field of microsimulation. This book provides a very useful summary of the state of the art in relation to spatial microsimulation. Along with other recent and new texts (O'Donaghue et al. 2012; Zaidi et al. 2009), we are beginning to provide the research community with a very impressive set of core texts.

Acknowledgements This chapter describes research that has been undertaken during the past 13 years at NATSEM, with funding assistance from a series of grants from the Australian Research Council and a significant number of government departments. We would like to acknowledge and thank the ARC (LP0349152, LP775396, LP0349126, DP664429) and our research partners: the NSW Department of Community Services; the Australian Bureau of Statistics; the ACT Chief Minister's Department; the Queensland Department of Premier and Cabinet; Queensland Treasury; the Victorian Departments of Education and Early Childhood and Planning and Community Development; the Australian Department of Health and Ageing; the NSW Department of Disability, Ageing and Home Care; the NSW Premier's Department and the Victorian Department of Sustainability and Environment. The views expressed in this chapter are those of the authors and cannot be interpreted as construing endorsement by any of the above agencies of any of the research methods or findings. We would also like to thank our fellow chief investigators on these grants and our international partner investigator, Dr. Paul Williamson. Thanks also to Yogi Vidyattama for assistance with the figure.

References

Ballas, D., Rossiter, D., Thomas, B., Clarke, G. P., & Dorling, D. (2005). *Geography matters: Simulating the local impacts of national social policies*. Joseph Rowntree Foundation: York.

Birkin, M., & Clarke, M. (1988). SYNTHESIS – A synthetic spatial information system for urban and regional analysis: Methods and examples. *Environment and Planning A, 20*(12), 1645–1671.

Birkin, M., & Clarke, M. (1989). The generation of individual and household incomes at the small area level using synthesis. *Regional Studies: The Journal of the Regional Studies Association, 23*(6), 535–548.

Birkin, M., & Clarke, G. P. (2011). The enhancement of spatial microsimulation models using geodemographics. *Annals of Regional Science*. doi: 10.1007/s00168-011-0472-2.

Caldwell, S. B., & Keister, L. A. (1996). Wealth in America: Family stock ownership and accumulation, 1960–1995. In G. P. Clarke (Ed.), *Microsimulation for urban and regional policy analysis*. London: Pion.

Cassells, R., Harding, A., Tanton, R., Miranti, R., & McNamara, J. (2010). *Spatial microsimulation: Preparation of sample survey and census data for SpatialMSM/08 and SpatialMSM/09* (NATSEM Technical Paper No. 36).

Chin, S. F., & Harding, A. (2006, April). *Regional dimensions: Creating synthetic small-area microdata and spatial microsimulation models* (Technical Paper No. 33). National Centre for Social and Economic Modelling.

Chin, S. F., Harding, A., Lloyd, R., McNamara, J., Phillips, B., & Vu, Q. (2005). Spatial microsimulation using synthetic small area estimates of income, tax and social security benefits. *Australasian Journal of Regional Studies, 11*(3), 303–336.

Clarke, M., & Holm, E. (1987). Microsimulation methods in spatial analysis and planning. *Geografiska Annaler Series B, 69*(2), 145–164.

Edwards, K. L., & Clarke, G. P. (2009). The design and validation of a spatial microsimulation model of obesogenic environments in Leeds: SimObesity. *Social Science & Medicine, 69*, 1127–1134.

Gong, C., McNamara, J., Vidyattama, Y., Miranti, R., Tanton, R., Harding, A., & Kendig, H. (2012). Developing spatial microsimulation estimates of small area advantage and disadvantage among older Australians. *Population, Space and Place, 18*(5), 551–565.

Gupta, A., & Harding, A. (Eds.). (2007). *Modelling our future: population ageing, health and aged care: International symposia in economic theory and econometrics*. Amsterdam: North Holland.

Gupta, A., & Kapur, V. (Eds.). (2000). *Microsimulation in government policy and forecasting*. Amsterdam: North Holland.

Harding, A. (Ed.). (1996). *Microsimulation and public policy* (Contributions to economic analysis series). Amsterdam: North Holland.

Harding, A., & Tanton, R. (2011). Policy and people at the small area level: using microsimulation to create synthetic spatial data. In R. Stimson (Ed.), *Handbook in spatially integrated social science research methods*. Sydney: Edward Elgar.

Harding, A., Hellwig, O., Bremner, K., & Robinson, M. (1999, December 2). *Geodemographics of the aged: Where they live, what they buy*. Paper presented at the 'Geodemographics of Ageing in Australia' Symposium, Brisbane, Australia.

Harding, A., Lloyd, R., Bill, A., & King, A. (2006). Assessing poverty and inequality at a detailed regional level: new advances in spatial microsimulation. In M. McGillivray & M. Clarke (Eds.), *Understanding human well-being* (pp. 239–261). Helsinki: United Nations University Press.

Harding, A., Vu, N. Q., Tanton, R., & Vidyattama, V. (2009). Improving work incentives and incomes for parents: The national and geographic impact of liberalising the family tax benefit income test. *The Economic Record, 85*, S48–S58.

Harding, A., Keegan, M., & Kelly, S. (2010). Validating a dynamic microsimulation model: Recent experience in Australia. *International Journal of Microsimulation, 3*(2), 46–64.

Harding, A., Vidyattama, Y., & Tanton, R. (2011). Demographic change and the needs-based planning of government services: Projecting small area populations using spatial microsimulation. *Journal of Population Research, 28*(2–3), 203–224.

Harland, K., Birkin, M., Heppenstall, A., & Smith, D. (2012). Creating realistic synthetic populations at varying spatial scales: A comparative critique of microsimulation techniques. *Journal of Artificial Societies and Social Simulation, 15* (1), http://jass.soc.surrey.ac.uk/15/1/1.html.

Lloyd, R. (2007). STINMOD: Use of a Static Microsimulation Model in the Policy Process in Australia. In A. Harding & A. Gupta (Eds.), *Modelling our future: Population ageing, social security and taxation, international symposia in economic theory and econometrics* (pp. 315–333). Amsterdam: North Holland.

Lymer, S., Brown, L., Harding, A., Yap, M., Chin, S. F., & Leicester, S. (2006). *Development of CareMod/05* (Technical Paper No. 32). Canberra: NATSEM, University of Canberra.

Lymer, S., Brown, L., Yap, M., & Harding, A. (2008a). Regional disability estimates for New South Wales in 2001 using spatial microsimulation. *Applied Spatial Analysis and Policy, 1*(2), 96–116.

Lymer, S., Brown, L., Harding, A., & Yap, M. (2008b). Predicting the need for aged care services at the small area level: The CAREMOD spatial microsimulation model. *International Journal of Microsimulation, 2*(2), 27–42.

McNamara, J., Tanton, R., & Phillips, B. (2007). *The regional impact of housing costs and assistance on financial disadvantage: Final report*. Australian Housing and Urban Research Institute: Melbourne.

Miranti, R., McNamara, J., Tanton, R., & Harding, A. (2011). Poverty at the local level: National and small area poverty estimates by family type for Australia in 2006. *Applied Spatial Analysis and Policy, 4*(3), 145–171. doi:10.1007/s12061-010-9049-1.

Mitton, L., Sutherland, H., & Weeks, M. (Eds.). (2000). *Microsimulation modelling for policy analysis*. Cambridge: Cambridge University Press.

Nakaya, T., Fotheringham, A. S., Hanoaka, K., Clarke, G. P., Balals, D., & Yano, K. (2007). Combining microsimulation and spatial interaction models for retail location analysis. *Journal of Geographical Systems, 4*, 345–369.

O'Donaghue, C., Ballas, D., Clarke, G. P., Hynes, S., Morrissey, K. L. (Eds.). (2012). *Microsimulation for rural policy analysis*. Berlin: Springer (forthcoming).

Rahman, A., Harding, A., Tanton, R., & Liu, S. (2010). Methodological issues in spatial microsimulation modelling for small area estimation. *International Journal of Microsimulation, 3*(2), 3–22.

Rao, J. N. K. (2002). Small area estimation: Update with appraisal. In N. Balakrishnan (Ed.), *Advances on methodological and applied aspects of probability and statistics* (pp. 113–139). New York: Taylor and Francis.

Rao, J. N. K. (2003). *Small area estimation*. Hoboken: Wiley.

Smith, D. M., Pearce, J. R., & Harland, K. (2011). Can a deterministic spatial microsimulation model provide reliable small-area estimates of health behaviours? An example of smoking prevalence in New Zealand. *Health & Place, 17*, 618–624.

Tanton, R., & Vidyattama, Y. (2011). Pushing it to the edge: extending generalised regression as a spatial microsimulation method. *International Journal of Microsimulation, 3*(2), 23–33.

Tanton, R., & Vu, Q. (2010). The distributional and regional impact of the Australian Government's household stimulus package. *Australasian Journal of Regional Studies, 16*(1), 127–145.

Tanton, R., Vidyattama, Y., McNamara, J., Vu, Q., & Harding, A. (2009a). Old, single and poor: Using microsimulation and microdata to analyse poverty and the impact of policy change among older Australians. *Economic Papers, 28*(2), 102–120.

Tanton, R., McNamara, J., Harding, A., & Morrison, T. (2009b). Rich suburbs, poor suburbs? Small area poverty estimates for Australia's eastern seaboard in 2006. In A. Zaidi, A. Harding, & P. Williamson (Eds.), *New frontiers in microsimulation modelling*. London: Ashgate.

Tanton, R., Harding, A., & McNamara, J. (2010). Urban and rural estimates of poverty: Recent advances in spatial microsimulation in Australia. *Geographical Research, 48*(1), 52–64.

Tanton, R., Vidyattama, Y., Nepal, B., & McNamara, J. (2011). Small area estimation using a reweighting algorithm. *Journal of the Royal Statistics Society Series A, 174*(4), 931–951.

Tomintz, M., Clarke, G., & Rigty, J. (2008). The geography of smoking in Leeds: Estimating individual smoking rates and the implications for the location for sleep smoking services,. *Area, 40*(3), 341–353.

Vidyattama, Y., Cassells, R., Harding, A., & Mcnamara, J. (2011, August 25). Rich or poor in retirement? A small area analysis of Australian private superannuation savings in 2006 using spatial microsimulation. *Regional Studies*. doi: 10.1080/00343404.2011.589829. Available online: http://www.tandfonline.com/doi/abs/10.1080/00343404.2011.589829)

Vidyattama, Y., & Tanton, R. (2010). Projecting small area statistics with Australian spatial microsimulation model (SpatialMSM). *Australian Journal of Regional Studies, 16*(1), 99–126.

Voas, D., & Williamson, P. (2000). An evaluation of the combinatorial optimisation approach to the creation of synthetic microdata. *International Journal of Population Geography, 6*, 349–366.

Williamson, P. (2001). *A comparison of synthetic reconstruction and combinatorial optimisation approaches to the creation of small-area microdata* (Working Paper 2001/2). Liverpool: Population Microdata Unit, Department of Geography, University of Liverpool.

Williamson, P., Birkin, M., & Rees, P. (1998). The estimation of population microdata by using data from small area statistics and samples of anonymised records. *Environment and Planning A, 30*(5), 785–816.

Wu, B., Birkin, M., & Rees, P. (2011). A dynamic MSM with agent elements for spatial demographic forecasting. *Social Science Computer Review, 29*, 145–160.

Zaidi, A., Harding, A., & Williamson, P. (2009). *New frontiers in microsimulation modelling*. Vienna: Ashgate.

Index

R. Tanton and K.L. Edwards (eds.), *Spatial Microsimulation: A Reference Guide for Users*, Understanding Population Trends and Processes 6, DOI 10.1007/978-94-007-4623-7,

The manufacturer's authorised representative in the EU is Springer
Nature Customer Service Centre GmbH, Europaplatz 3, 69115 Heidelberg,
Germany. If you have any concerns regarding our products, please
contact ProductSafety@springernature.com

Printed and bound by CPI Group (UK) Ltd, Croydon, CR0 4YY
30/09/2026
02225508-0009